Knowing and Acting
in Medicine

Knowing and Acting in Medicine

Edited by
Robyn Bluhm

ROWMAN &
LITTLEFIELD
INTERNATIONAL

London • New York

Published by Rowman & Littlefield International, Ltd.
Unit A, Whitacre Mews, 26-34 Stannary Street, London SE11 4AB
www.rowmaninternational.com

Rowman & Littlefield International, Ltd. is an affiliate of Rowman & Littlefield
4501 Forbes Boulevard, Suite 200, Lanham, Maryland 20706, USA
With additional offices in Boulder, New York, Toronto (Canada), and Plymouth (UK)
www.rowman.com

British Library Cataloguing in Publication Data
A catalogue record for this book is available from the British Library

ISBN: HB 978-1-7834-8809-4
      PB 978-1-7834-8810-0

Library of Congress Cataloging-in-Publication Data is Available
ISBN 978-1-78348-809-4 (cloth: alk. paper)
ISBN 978-1-78348-810-0 (pbk: alk. paper)
ISBN 978-1-78348-811-7 (electronic)

♾ ™ The paper used in this publication meets the minimum requirements of American
National Standard for Information Sciences—Permanence of Paper for Printed Library
Materials, ANSI/NISO Z39.48-1992.

Printed in the United States of America

Contents

Introduction

Robyn Bluhm

Medicine raises a number of interesting issues for philosophy and is also an area where philosophy has much to contribute. While the obvious example of a productive exchange between medicine and philosophy is bioethics, philosophers of medicine are increasingly writing about epistemological questions, as well, in part because they recognize the extent to which epistemological and ethical questions are interrelated. Ethical concerns shape the way we seek and use knowledge, and conversely, both what counts as knowledge in medicine and the way we use that knowledge have ethical implications.

The chapters in this volume all examine the relationship between knowledge and action in medical practice, research, or policy. "Knowledge" is broadly understood to include knowledge from clinical, laboratory, or social science research, and from the clinical encounter, as well as broader background assumptions prevalent in society that inform both the kinds of knowledge that are taken to be relevant to medicine and how that knowledge is interpreted in decision-making. The chapters cover a wide variety of topics and address distinct aspects of medicine, but despite this variety, a number of themes recur throughout the book. In this introduction, I identify four such themes: investigating how to understand and evaluate clinical reasoning; attending to how patients, and the public more broadly, understand and can contribute to medical knowledge; identifying the philosophical assumptions that underlie influential movements in medicine; and clarifying key concepts and/or goals in medicine.

Clinical reasoning is currently a topic of great interest in medicine itself, largely because of the influence of evidence-based medicine (EBM). From its first introduction in the early 1990s, EBM has aimed to teach clinicians to find, assess, and use the best available research evidence in making decisions regarding the care of individual patients. Yet although EBM recognizes that the results of clinical trials, which it views as the highest form of evidence, must be "integrated" with other information in order to be useful in the clinic, in practice it still offers almost no guidance as to how to do this. The chapters by Mark Tonelli (chapter 1) and by Sophie van Baalen and Mieke Boon (chapter 2) both offer ways of thinking about clinical reasoning that overcome this shortcoming of EBM. Tonelli draws on Stephen Toulmin's work in argumentation theory to develop an account of clinical reasoning based on warrants. Unlike EBM's limited conception of evidence, warrants can come from a variety of sources, and no source takes precedent over others (or is said to provide a "base" for decisionmaking).

van Baalen and Boon begin their account of clinical reasoning with work in philosophy of science that looks at the use of models in scientific research and extend this to the kind of epistemic work that doctors undertake in clinical practice. Their approach also aims to account for the frequent claim that much of doctors' reasoning is tacit, but to avoid allowing tacit knowledge to become mysterious and unknowable, so that doctors can be expected to make their knowledge explicit, if necessary.

Another chapter in which clinicians' reasoning processes is central is Anthony Fernandez's examination of diagnosis in psychiatry (chapter 3). Fernandez draws on phenomenological analyses of diagnostic reasoning, introducing the influential work of Osborne Wiggins and Michael Schwartz and addressing some common misunderstandings or misinterpretations of their approach. He shows that if we conceive of psychiatric classification and diagnosis in terms of "ideal types," we can both do descriptive justice to what psychiatrists actually do in the clinic and understand mental disorders in a way that is robust enough to guide current practice, but flexible enough to be adapted to new scientific knowledge.

Yet, of course, no matter how good clinical reasoning may be, it is still fallible. Abraham Schwab (chapter 4) builds an account of epistemic humility that takes into account both the need to avoid overconfidence and the fallible nature of the scientific evidence that informs clinical decisions. Like van Baalen and Boon, Schwab uses work in philosophy of science to develop his argument; clinicians' degree of confidence in scientific results should depend on the extent to which they are reproducible and, in a sense developed by William Wimsatt and by J.D. Trout, robust. As with all of the chapters that look at clinical reasoning, Schwab's chapter emphasizes that good reasoning requires both the epistemic skills to critically assess information from a variety of sources and the ethical motivation to critically assess one's own reasoning process.

Medical knowledge is not, however, something that only clinicians possess. A number of movements in medicine, such as patient-centered medicine and values-based practice, have aimed to better incorporate the perspective of those who use health care services. This question is also addressed by several chapters. Tania Gergel and Thomas Kabir (chapter 5) survey the growing movement in the UK to conduct research on mental health conditions that includes service users. While great progress has been made in including users' perspectives at all stages of research (planning, data collection and analysis, and dissemination of results), Gergel and Kabir also identify remaining challenges.

Rachel Ankeny (chapter 6) also focuses on the importance of including patient perspectives in medical research, specifically looking at case studies. Her analysis identifies the epistemological and ethical values reflected in the use of patient perspectives in published case reports, and she concludes that this practice provides a rich source of evidence.

Maya Goldenberg and Christopher McCron (chapter 7) also address issues regarding medical knowledge and the public, examining media uptake of a study that examined the effects of a pro-vaccine message on parents' reported plans to vaccinate their children. The study, by Brendan Nyhan, found that their intervention had a "backfire" effect on some parents who were hesitant about vaccination prior to the study. Goldenberg and McCron show that the media's reporting of the study was often inaccurate and overblown, and they identify some of the possible repercussions of this inaccuracy for public discussion of vaccination.

Like Goldenberg and McCron's contribution, Susan Hawthorne's chapter (chapter 8) involves a careful textual analysis and the identification of ethical and epistemological issues raised by her sources. Specifically, Hawthorne examines the evidence that was available in the United States in the early 1960s for the deinstitutionalization of people suffering from serious mental disorders, in favor of community-based care. She shows that, while there was plenty of evidence that the current situation was deeply problematic, there was not enough evidence to support a particular alternative. Instead, she demonstrates, the change was supported primarily by ethical and social values. But this does not mean that the decision should have been value free. Rather, Hawthorne argues that values and evidence are always interrelated in policy decisions; the appropriate response is to aim to have both strong evidence and well-justified values.

Just as Hawthorne questions the basis for mental health policy, Keekok Lee (chapter 9), Delphine Olivier (chapter 10), and Marianne Boenink (chapter 11) question contemporary understandings of medical knowledge. Lee examines the currently dominant framework of medical knowledge provided by EBM, with its reliance on randomized controlled trials (RCTs), identifying several assumptions on which this framework depends and arguing that these assumptions actually threaten to undermine the knowledge that they purportedly support.

Olivier's chapter focuses on another aspect of medicine that is often assumed to be well justified, that is, preventive medicine. She shows that, even in cases where we know that there is risk of disease, and have the interventions to reduce that risk, preventive medicine may not be completely benign.

Similarly, Marianne Boenink (chapter 11) shows that a variety of meanings have been attached to the concept of "translational medicine," showing that where one locates, and how one characterizes, the "gap" between research and health outcomes is important. Yet all of these characterizations, she argues, aim to solve a problem that is a symptom of a broader underlying cause; drawing on work in the history of medicine, she suggests that, in thinking about the way that research can help improve clinical care, we must also think about how our concepts of disease and of health shape biomedicine and *create* the gap that translational research is supposed to close.

While the chapters I have just described all look at "large scale" aspects of medical knowledge, other chapters focus more narrowly on a specific concept. Suze Berkhout (chapter 12) draws on her empirical research working with women living with HIV/AIDS to consider the multiple meanings of "adherence." She shows that patients, nurses, and physicians may have quite different understandings of what it means for a patient to adhere to her regimen of antiviral medication, and she uses the literature on the epistemology of ignorance to consider some of the possible results of these incompatible understandings.

Continuing the theme of careful attention to the way concepts are used in medicine, James Krueger (chapter 13) looks at the way that the distinction between "treatment" and "supportive care" was made in the recent Ebola outbreak. He identifies two distinct, and problematic, ways of drawing the distinction and then offers an alternative approach; he also draws out some of the practical reasons for attending to what might seem like a purely terminological issue.

Finally, Joseph Goode's discussion in chapter 14 of the use of simulation in health care education shows that philosophy can have practical outcomes. Goode uses Dewey's pragmatism to critically assess the theoretical rationale for the use of simulation and shows that much of the current literature does not take full advantage of the resources provided by Dewey's approach and may use concepts that are central to Dewey's philosophy, but in ways that do not reflect Dewey's thinking as a whole. As a result, educational programs may not benefit from the advantages a Deweyan theory can provide.

Chapter 1

Case-based (Casuistic) Decision Making

Mark R. Tonelli

While medicine encompasses, incorporates, and is influenced by a range of disciplines, from basic science research to politics, from public health to economics, the practice of medicine remains focused on the provision of care aimed at benefiting individuals. Clinical medicine remains a personal undertaking, relating one individual in need of healing with another who professes and promises to heal (Pellegrino and Thomasma 1988). In this effort to provide care to individual patients, clinicians must seek to incorporate all relevant knowledge into the decision-making process.

Knowledge in medicine serves primarily as a guide to clinicians endeavoring to benefit individual patients. Medical knowledge utilized by clinicians can be divided into three topic areas: that derived from clinical research, that stemming from pathophysiologic understanding, and that based on clinical experience. Each of these topics produces and/or provides backing for warrants with particular strengths and weaknesses with regard to clinical decision making (Tonelli 2006).

The process of clinical reasoning closely resembles the model of argumentation described by Stephen Toulmin (2003), where warrants supporting a decision must be elucidated, examined, and weighed against one another. Clinicians begin by establishing the warrants that are relevant to the decision at hand. When warrants are consistent, for instance, when a randomized-controlled trial, the underlying pathophysiology, and a clinician's experience all support the use of a particular medication for a particular ailment, clinical judgment is straightforward. Clinicians must also incorporate other warrants derived from patient experiences, goals, and values as well as features of the medical delivery system itself into a final decision or recommendation.

When warrants derived from medical knowledge conflict, clinicians must examine the backing and weigh the relative strength of individual warrants.

As the knowledge derived from each of the topics differs in kind, not degree, purported hierarchies of knowledge or evidence that seek to include all are epistemically untenable. In any particular case, the most compelling warrants may derive from any of the topic areas. Clinical reasoning must be based on the case at hand, a casuistic endeavor rather than a deductive process. As such, conclusions are only probable and remain subject to rebuttal. Clinicians who agree on the medical knowledge relevant to a clinical decision may disagree on the decision itself due to differential weighting of warrants. Patient-derived and system-derived warrants will ultimately need to be incorporated into a final decision or recommendation.

INDIVIDUALIZED MEDICINE

In seeking to arrive at the best decision or recommendation for an individual patient, clinicians will tend to focus their attention on a limited subset of medical knowledge. An understanding of medical history helps to clarify the major categories of medical knowledge. In the earliest forms of Western clinical medicine, the experience of the clinician was paramount. At a time when effective therapeutic interventions were scarce, the experience of a clinician aided accurate diagnosis, thereby improving prognostication, often the only real value a physician could add to the care of the patient. Trial and error were a mainstay of therapy. Yet even early in the recorded history of medicine, physicians sought to move beyond this dependence on experiential empiricism toward a more rational approach to diagnosis and treatment. Rationalism required understanding the nature of illness in order to arrive a diagnosis and/or treatment plan. That medicine's understanding of the nature of illness remained quite poor for nearly two millennia certainly led to ineffective or harmful interventions, such as bleeding and purging, that were otherwise perfectly rational. Since the mid-twentieth century, the focus of medical epistemology has shifted to a new empiricism, the use of tools designed to improve public health and to elucidate direct causation for generating knowledge relevant to clinical decisions. Empirical approaches have become particularly privileged in the realm of evaluating the direct, causative effects of new therapeutic interventions, where the mechanistic approach of pathophysiologic reasoning seemed to be particularly fallible (Howick 2011). The inclination and ability to measure outcomes allow for a more mathematical demonstration of causality seemingly not as subject to the errors caused by the incomplete and imperfect understanding of mechanism. The new empiricism found its voice and established its prominence with the development and evolution of the evidence-based medicine (EBM) movement (Evidence-based Medicine Working Group 1992). But while rationalism once largely eclipsed experiential practice and the new empiricism has sought to replace rationalism, experience and pathophysiologic reasoning remain part of clinical practice.

Clinical judgment requires the incorporation of multiple kinds of medical knowledge as well as patient-specific and contextual features. The three topic areas of medical knowledge are (1) experiential knowledge, (2) pathophysiologic (also referred to as mechanistic) reasoning, and (3) the results of clinical research. These topic areas are advanced as mutually exclusive and exhaustive (with regard to clinically relevant medical knowledge). Medical knowledge, however, is not all that needs to be considered by clinicians when arriving at a clinical judgment. A clinician must certainly elicit and incorporate the individual patient's goals, values, and experiences into recommendations and decisions. Similarly, the system (economic, political, cultural, legal) in which medicine is practiced will provide inducements and constraints upon a clinician's and patient's choices independent of medical knowledge. The primary focus of this discussion, however, will be on the value and the incorporation of the three kinds of medical knowledge into clinical decision making.

The scientific basis of evidence-based medicine, epidemiology, has its origins in public health, evolving to include controlled trials in order to more convincingly demonstrate causation (Pearce 1996). But while the notion of clinical practice as clinical epidemiology has gained acceptance with the advancement of the EBM movement, the limitations of this kind of medical knowledge for clinicians have also become clearer. The direct use of the results of epidemiology and controlled trials, results that reflect effects across populations, would seem likely to improve public health. The application of such results to individual cases is more problematic (Tonelli 1998; Naylor 1995). If the goal of clinical medicine remains defined as the attempt to improve the care of individuals, then general knowledge obtained from clinical research will only go so far. Utilizing experience and pathophysiologic reasoning will remain necessary to reach the best decision for an individual patient (Tonelli 2010).

STRENGTHS AND WEAKNESSES OF THE
KINDS OF MEDICAL KNOWLEDGE

A great deal of attention over the last two decades has focused upon the value of the results of clinical research for clinical decision making. Broadly speaking, the results of clinical research provide clinicians with medical knowledge that is derived in a way that minimizes bias, can be subject to careful scrutiny, and is relatively easy to disseminate. A well-designed observational or controlled study has the ability to discern small, but clinically relevant effects that no amount of individual experience would ever be able to. These strengths are important, and clinical research results often inform clinical decision making and patient choice (Tonelli 2010).

The limitations of the results of clinical research for clinical decision making, however, have tended to get short shrift in the literature of EBM. The primary weakness of clinical research is that it has no direct applicability to individual patients. That is, the results of any population study tell us little about how the patient at hand will respond. At best, clinical research tells us something about an "average" patient, but no patient is average and clinicians should always be vigilant regarding how those individual differences affect the relevance of clinical research results (Feinstein and Horwitz 1997). Additionally, clinical research results are hardly free from bias, particularly when it comes to how they are funded (Lexchin et al. 2003). Research results are certainly fallible and, at times, untrustworthy (Upshur 2009). Even when well done, research is fixed in a time and place, while clinical medicine is context dependent and ever evolving.

While EBM has sought to deemphasize pathophysiologic reasoning in clinical medicine, finding it unreliable, biological plausibility and physiologic understanding remain crucial to the development of medical knowledge and the practice of clinical medicine. The limitations of mechanistic reasoning are very real (Howick 2011). Physiology does not always predict relevant outcomes, as demonstrated in a study that noted decreased arrhythmias but increased mortality in post-myocardial infarction patients treated with antiarrhythmics (CAST Investigators 1989). And certainly the value of pathophysiologic reasoning depends on a firm understanding of the mechanisms of disease. But interestingly, our reliance on such reasoning appears to be decreasing at a time when our understanding of disease mechanisms is accelerating.

In clinical practice, incorporating the physiologic aspects of a patient's presentation or response to therapy allows for the importance and incorporation of individual variability. How individual patients differ from the "average" patient of a clinical trial generally comes down to psychological and physiological features, including comorbidities and specific manifestations of illness. Clinicians also monitor the physiologic responses of patients to facilitate early recognition of therapeutic value or futility. In the realm of clinical research, appeals to biologic plausibility can help to provide a check on spurious findings, for instance, studies suggesting the benefit of homeopathic remedies or the value of retroactive intercessory prayer (Giacomini 2009).

As with pathophysiologic reasoning, EBM has also focused on the limitations of relying on unsystematic clinical experience, personal or pooled, as a source of medical knowledge. Certainly, clinicians as persons are subject to a wide variety of cognitive biases (Kempanien et al. 2003). In individual practice, we often see only a small number of patients with a particular problem, making assessment of response to and risk from interventions difficult. Experience tends to lead to static practice patterns, with older physicians slower to incorporate new knowledge from clinical research (Choudhry et al. 2005).

Yet, clinical experience remains vital to individualizing care (Malterud 1995; Tonelli 1999). Experience, like pathophysiologic reasoning, allows clinicians to incorporate specific aspects of an individual's presentation into a clinical decision. In diagnostics, where pattern recognition serves as a foundational skill, experience increases accuracy and efficiency (Norman 2007). For clinicians who see a large number of patients with a specific disorder, clinical expertise provides them with a rich set of cases to which a new patient can be compared. This kind of expert knowledge, founded on clinical experience, can be transmitted to others. Finally, clinical experience generally provides the first indication of change in medicine, from the appearance of new disorders to changing manifestations of old ones. An example from my own experience, the sudden appearance of multiple cases of *Mycobacterium abscessus* infection in our cystic fibrosis population, led us to be concerned about the possibility of person-to-person transfer of this infection. The medical literature at the time noted that there was no evidence of any person-to-person transfer of nontuberculous mycobacteria of any kind. Trusting our experience, we investigated further and determined that our new cases were due to the same genetic strain of the mycobacteria, strongly suggesting person-to-person spread. A change in infection control practices had been instituted even prior to this conclusion, based on our experience, and no new cases appeared subsequently.

In summary, clinically relevant medical knowledge comprises three different kinds of medical knowledge, none of which deserves general priority when it comes to clinical decision making. Each kind of medical knowledge has strengths and weaknesses with regard to its applicability to clinical reasoning. A clinician, then, must bring all relevant knowledge to bear upon a particular case if he or she hopes to make what is probably the best choice for that individual patient.

ELUCIDATING WARRANTS

Clinical judgment requires elucidating and potentially negotiating between multiple warrants. Warrants represent a reason to support a particular judgment (Toulmin 2003) and in medicine are often derived from and backed by clinical research, pathophysiologic reasoning, and clinical experience. Warrants generally contain both factual and value-laden elements, and may be backed by a variety of factual and/or value-laden claims. In this way warrants differ from "evidence" as currently defined by EBM. For instance, invoking a warrant such as "use of a beta-blocker in persons who have had a recent myocardial infarction will decrease the risk of death" may be backed by relevant clinical research and pathophysiologic understanding, but must contain a value element along the lines of "decreasing the risk of death is a good thing" in order to be used to support a decision for a particular patient. As with epistemic backing, the strength of the ethical backing for a particular warrant will vary. In particular, the clinician will need to carefully elucidate the experiences, goals, and values of the individual patient in order to properly weigh warrants. For our example of a reduction in risk of death, the normative component of a warrant may seem indisputable, yet this goal is often not appropriate for patients whose quality of life is already below an acceptable minimum. The value-laden components of warrants need to be made explicit and examined. While we can assume that most patients would like to reduce their risk of death, not all see that as a major benefit. Often, outcomes that are easily measured (e.g., reduced cholesterol) and frequently utilized in clinical research are not by themselves seen as important by most patients. The recognition of this normative gap between clinical research results and patient values has led to a movement to encourage clinical researchers to focus on patient-centered outcomes (Selby et al. 2012). In doing so, researchers would be strengthening the backing for clinical warrants derived from their work.

Each relevant warrant, then, provides support for an argument regarding the best course of action for an individual patient. For instance, we may determine that the person before us complaining of a nonproductive cough, nasal congestion, and a sore throat has a viral respiratory tract infection. When the patient raises the possibility of antibiotics for this infection, clinicians are likely to (and should) invoke the warrant that "antibiotics are not appropriate treatment for viral upper respiratory tract infections (URIs)" (Prendergast 1995). If the patient (or a medical trainee) is not convinced of the warrant on its face, then backing for the warrant can be provided. Warrants utilized in clinical medicine will generally be backed by knowledge derived from clinical research, pathophysiologic reasoning, or clinical experience. In this particular case, each of the three kinds of knowledge would provide backing for the warrant: clinical research does not demonstrate a benefit of antibiotics in persons with viral URIs; understanding how antibiotics work leads us to believe that they will have no effect on viral replication; experience has demonstrated that such infections are self-limited and do not get better more quickly with antibiotic therapy. The value-laden (and generally unspoken) component of the warrant is a claim that clinicians ought not provide interventions that have no reasonable chance of providing benefit, particularly those that have some attendant risk, even if requested by patients. The therapeutic judgment here is relatively straightforward, though certainly the conclusion rests upon how certain we are with our initial diagnosis.

Elucidating and negotiating between warrants could be approached in several manners. A theoretical construct, one that attempts to set a pre-determined hierarchy of warrants, based on the type or backing, could be considered. Evidence-based medicine appears to represent such an approach, with a persistent assertion of and adherence to a hierarchy of "evidence" that is actually a hierarchy of research methodologies. The theoretical preference for knowledge derived from particular methodologies could be used to negotiate between conflicting warrants. In such a model, the warrant backed by the "best evidence" (i.e., the most rigorous study design) wins the argument and determines the course of action. The multiple problems with such an approach have been elucidated in convincing critiques of EBM and will not be revisited here (Tanenbaum 1993; Feinstein and Horwitz 1997; Bluhm 2005; Tonelli 2006; Goldenberg 2006).

Another approach to negotiating between conflicting warrants would be to focus on their relevance and weight for each particular case. This casuistic approach acknowledges that clinical medicine is a practical discipline, one that must utilize both theory and experience. While casuistic reasoning in medicine has been primarily advanced as a methodology to deal with the ethical issues involved in difficult clinical cases (Jonsen et al. 2002), there is no reason to limit the method to ethics only. Jonsen and Toulmin, in *The Abuse of Casuistry*, use clinical medicine as a paradigmatic example of a casuistic discipline, noting that "no professional enterprise today is closer to moral practice, or better exemplifies the special character of 'practical' inquiries, than clinical medicine" (Jonsen and Toulmin 1988, 36). A casuistic approach to clinical decision making offers the advantage of doing both the medical and the ethical work simultaneously, avoiding the "two-legged" approach advocated by some (Fulford et al. 2012).

From a practical standpoint, clinicians can use the framework of three topic areas of clinical knowledge to ensure that they are elucidating all of the relevant warrants for a particular clinical question. Not stopping with an appeal to clinical research alone, a clinician must at least consider whether physiologic or experiential warrants come into play. The challenges in clinical judgment arise when these various warrants conflict.

CLINICAL JUDGMENT AND CONFLICTING WARRANTS

Conflicting warrants require a clinician to weight the importance of any particular warrant for the patient at hand. As with any casuistic endeavor, there are no hard and fast rules that dictate how clinicians deal with conflicting warrants; casuistry rejects the notion that any single type of warrant should be considered to have general priority. This is where a case-based approach most obviously conflicts with the approach advanced by EBM. On EBM hierarchies that include pathophysiologic rationale and/or experiential knowledge, knowledge derived from clinical research, even poor quality clinical research, is always advanced as superior. This hierarchical view of medical knowledge cannot be defended in clinical practice, where cases abound where sound clinical judgment will mean that warrants based on physiologic reasoning or experience will trump those derived from clinical research when it comes to choosing the best course of action for a particular patient.

For example, in critical care medicine a body of evidence derived from clinical research supports the notion that using lower set tidal volumes on ventilated patients with acute respiratory distress syndrome (ARDS) results in lower hospital mortality when compared with high tidal volumes. Yet, there are individual patients who do not do well physiologically when being ventilated with the tidal volume (6 ml/kg ideal body weight) used in the most compelling clinical study. I have had patients who developed life-threatening cardiac arrhythmias or refractory hypoxemia not responsive to other measures when their tidal volumes were turned down. In these cases, the life-threatening physiology can often be ameliorated by a slight increase in the tidal volume delivered. To insist that the warrant derived from clinical research, that a 6 ml/kg tidal volume must be adhered to regardless of the physiologic response of the patient, would seem to represent poor clinical judgment. Such a preference for a warrant backed by clinical research over a conflicting warrant backed by pathophysiologic reasoning errs by both failing to recognize that no type of medical knowledge has general priority for clinical decision making and in not recognizing the limitation of medical knowledge derived from clinical research. In the latter sense, the most relevant study, for reasons of experimental expediency, compared only 6 ml/kg to 12 ml/kg, not to the 8 ml/kg it might take to reverse the life-threatening physiologic aberrations in the patients discussed above. Clinical research really provides very little backing to the notion that 6 ml/kg is better for ARDS patients than 8 ml/kg, severely weakening the strength of the warrant that 6 ml/kg will improve survival in those patients who tolerate it poorly from a physiologic standpoint. Thoughtful proponents of EBM generally acknowledge that clinical research should not dictate care, but beyond calling for clinicians to "integrate" clinical research into practice, little guidance is provided (Sackett et al. 1996). Specifically, the ongoing defense of a hierarchy of evidence in EBM makes it difficult for clinicians to defend elevating pathophysiologic reasoning or experiential knowledge over clinical research (Djulbegovic et al. 2009; Howick 2011).

Clinicians will also weigh the importance of warrants based on clinical research differently from one another for other reasons. First, clinicians will evaluate any piece or body of clinical research through a lens of prior probability. That is, clinicians are Bayesian thinkers who will hold a pretest opinion regarding the likely value of a diagnostic tool or therapeutic option prior to assessing a new piece of clinical research into that intervention. These priors may be the result of unrecognized cognitive biases, but they may also represent a well-considered position based on a combination of previous clinical research, pathophysiologic understanding, and/or clinical experience (Rubenfeld 2001). Clinicians with different priors may look at the same products of clinical research and reach different conclusions regarding the applicability of the results to a particular patient or to clinical practice in general. (The tendency toward strong priors among clinicians leads to the quip that there are only two responses to any report of clinical research, either "I already knew that" or "I do not believe that.")

Clinicians may also find the backing for warrants derived from clinical research weaker or stronger on the basis of factors not related to the strength of the evidence, per se. That is, while proponents of EBM define the strength of evidence on the basis of study design and statistical robustness, clinicians will find that other features of a particular study make the results more or less compelling for clinical practice (Tonelli 2012). These factors can be divided into three groups. First, epistemic considerations include prior belief, biologic plausibility, consistency, and objectivity of research findings. Second, in hoping to provide benefit to individual patients, clinicians will be concerned about the applicability, safety, time to effect, effect size, and the value of the outcome to patients. Finally, in order to exercise the duty of stewardship, clinicians will also need to be concerned about the cost and ease of implementation of an intervention, as well as the availability, cost, and efficacy of any alternative treatments. Consideration of these factors, which may be done differently by different clinicians, will affect the weight given to a warrant derived from clinical research as it is applied to an individual patient.

Ultimately, clinicians will need to not only negotiate between potentially conflicting warrants derived from and backed by various kinds of medical knowledge but also incorporate warrants derived from patient experiences, goals, and values as well as from the realities of the system in which clinical medicine is practiced. For instance, a recommendation that requires inpatient treatment will likely not be appropriate for a terminally ill patient who values time at home with family. Similarly, studies that demonstrate the value of highly technological interventions, no matter how rigorous, will not provide a meaningful warrant in a system where the technology is unavailable.

When negotiating between warrants in clinical practice, casuistic reasoning results in only probabilistic conclusions. Clinicians work in a realm with a high degree of uncertainty, so it is no surprise that their judgments will remain only probably right and are always open to rebuttal. Rebuttal may come from colleagues or from patients and family members. Rebuttals may take the form of a new, previously unconsidered warrant, may challenge the backing of an existing warrant, or may be a call for reweighting warrants already elucidated. Clinical judgment, then, should be able to be made explicit. Being able to clearly state what warrants are being invoked in a clinical case and how those warrants are being weighted allows for others to provide a check on the reasoning of individual clinicians. While there may be tacit elements important to clinical practice (Henry 2006), clinicians can still be explicit even if just to say they are relying on intuition.

Casuistic clinical reasoning often relies on analogy. Does the patient at hand closely resemble a paradigmatic case, a case where there is broad agreement on the appropriate course of treatment? Paradigmatic cases themselves may derive from clinical research (the "average" patient enrolled in a clinical trial) or from clinical experience. Differences of many kinds will serve to separate the case at hand from the paradigmatic case, with greater distance leading to greater uncertainty in the correctness of the clinical judgment, until the case at hand begins to more closely resemble another paradigmatic case with a different treatment plan. In these circumstances, the relative weighting of warrants shifts as the patient at hand differs from the paradigmatic case.

Considering the three kinds of medical knowledge and incorporating patient experiences, goals and preferences as well as system features provide a practical framework to aid clinicians in reaching a sound clinical decision.

Case-based Clinical Reasoning: An Example

In their article "The End of the Disease Era," Tinetti and Fried (2004) call for the abandonment of the clinical focus on disease and the adoption of an integrated, individually tailored model of clinical care. Casuistic clinical reasoning represents an ideal approach to achieving individualized care. One of the worked clinical examples offered by Tinetti and Fried is reconsidered here to demonstrate how a casuistic approach yields an individualized plan of care. Not surprisingly, a casuistic analysis yields treatment recommendations nearly identical to the approach deemphasizing disease. In fact, I would suggest that Tinetti and Fried employed case-based reasoning in reaching their conclusions.

> Mary is a 76-year-old woman with multiple medical problems, including congestive heart failure, diabetes, hypertension, atrial fibrillation and depression. While she is widowed and lives alone, she actively participates in the care of her three grandchildren living nearby.
> Mary presents to her primary care physician feeling more fatigued and less able to cope with her day-to-day life. She is eating less, primarily because she lacks an appetite and the energy to cook for herself. But she also does not like the food her low-sodium, "Heart Healthy" diet prescribes for her. She has been losing weight. (Tinetti and Fried 2004)

Mary wants to feel better and understands not all of her symptoms are attributable to her medical illnesses. She has great faith in her primary care physician and has always been very adherent to her prescribed medical regimen, but she is beginning to think that her medical therapies may be making her feel worse.

Establishing warrants:
Clinical research

- Multiple pharmaceutical interventions have demonstrated survival benefit for patients with heart failure, diabetes, hypertension, and atrial fibrillation.
- Evidence-based clinical practice guidelines recommend multiple pharmacologic interventions and dietary restrictions for each of these conditions.

Pathophysiologic reasoning

- Increasing number of medications increases the likelihood of drug interactions and of adverse effects.
- Given the multiple medical conditions being treated, it is not clear that following the guidelines for each disease is necessarily better for Mary, as most studies exclude patients with comorbid conditions.

Clinical experience

- Fatigue is generally a symptom with multifactorial causes.
- Mary states that she has generally felt worse with each additional medication that has been added to her regimen.
- Her physician knows that Mary will do whatever is recommended for her.

Patient goals and values

- Mary wants to feel better in her day-to-day life, which is more important to her than longevity.
- Mary feels the need to be useful, which includes being able to help with her grandchildren and volunteer at her church.

System features

- Mary has both Medicare and a supplemental plan, meaning her medication costs are largely covered.

• The quality of care provided by the physician is evaluated, in large part, by whether or not her patients are receiving therapies according to clinical practice guidelines. Some of the physician's remuneration is tied to these quality metrics.

Negotiating among warrants:
In an effort to provide the medical care demonstrated to increase longevity and decrease untoward events, Mary's physician has attempted to adhere to multiple clinical practice guidelines. The physician also has a financial incentive to adhere to these guidelines, because part of her compensation is based on the percentage of her patients who are being treated in accordance with relevant guidelines. Mary's fatigue and functional decline, however, are more likely to be the result of her polypharmacy than attributable to the diseases themselves. In carefully examining her goals, it is clear that Mary wants to remain active and engaged, even if that comes at the cost of some increased risk of medical complications or even death.

In weighting the many warrants relevant to Mary's care, those backed by clinical research demonstrating survival benefit become less compelling given Mary's focus on improving her quality of life. Pathophysiologic and experiential warrants related to the unpredictable and untoward effects of polypharmacy are given increased weight. Mary's physician must reevaluate her treatment regimen, with an eye to eliminating prescriptions and proscriptions that are likely only of marginal benefit. Mary, with her multiple comorbidities, differs significantly from the subjects of the clinical studies that gave rise to the practice guidelines now being used to dictate her care, meaning that the "average" patient in these studies differs significantly from Mary. Given Mary's deference to clinicians, she will need to be given explicit permission to liberalize her diet and to discontinue specific medications that are most likely to be contributing to her symptoms (e.g., beta-blockers). She should be encouraged and empowered to trial herself off some medications to assess the effect on her well-being. The physician must recognize that deviating from clinical practice guidelines is best for Mary, even if it comes at a cost for the clinician.

Shared Decision Making

Mary's experience with illness as well as her goals and preferences regarding treatment are incorporated into the clinician's medical decision making. The clinician does not simply develop a menu of possible therapeutic options, presenting such a list to Mary and allowing her to choose. This distinction is crucial, yet neither EBM nor patient-centered medicine, which does attempt to individualize care, seems to have envisioned this form of "shared decision making."

While shared decision making has been endorsed by evidence-based practitioners and their critics alike, what actually constitutes shared decision making remains nebulous (Moumjid et al. 2007). Shared decision making is often contrasted with paternalism, the widely discredited approach to clinical care where the clinician determines what is in the best interest of the patient with minimal or no input from the patient (Emanuel and Emanuel 1992). Any notion of shared decision making, however, must involve an exchange of information between clinician and patient (Charles et al. 1997). The basic view of this information exchange has clinicians providing medical information (facts) in a morally neutral manner to patients, with patients sharing their goals and preferences (values) (Brock 1991).

The simplistic notion that shared decision making incorporates facts provided by the clinician and values provided by the patient only serves to demean the physician and patient alike, leading to an inevitable and unnecessary conflict between facts and values, clinicians and patients. It is the duty of the clinician to elicit relevant patient experiences, preferences, and values and then to incorporate these factors while elucidating and weighing the relevant warrants for a particular clinical decision. The decision of the clinician may represent a specific recommendation to a patient or a specific action taken by the clinician, depending on the circumstances. But the process of clinical decision making remains intrinsic to the clinician. In the casuistic model, individual patients or their surrogates maintain the option to rebut the assumptions or reasoning of their clinicians. Such a rebuttal necessitates reconsideration of the clinical judgment by the clinician.

SUMMARY

Clinical medicine, if directed toward the benefit of individual patients, must remain more than applied clinical epidemiology. While clinical research certainly informs clinical practice, so does medical knowledge derived from pathophysiologic understanding and from direct clinical experience. Patient experience, goals, and values must be incorporated into medical decisions, not considered separately from them. The society, culture, and system in which medicine is practiced may facilitate or produce barriers to delivering certain kinds of care.

To care for an individual, the clinician must maintain focus on the individual. The patient does not generate a clinical question, a question that is answered by the appropriate acquisition, evaluation, and incorporation of clinical research results. The clinical question is always the same: What is the best thing to do for this particular patient? Answering this requires a careful examination and evaluation of what it is that makes the patient particular. Casuistry is rooted in the particulars of the case. To treat the patient as an individual, then, requires casuistic reasoning.

REFERENCES

Aitken, Moira L, Ajit Limaye, Paul Pottinger, Estella Whimbey, Christopher H. Goss, Mark R. Tonelli et al. 2012. "Respiratory Outbreak of Mycobacterium Abscessus Subspecies Massiliense in a Lung Transplant and Cystic Fibrosis Center." *American Journal of Respiratory and Critical Care Medicine* 185(2):231–2.

Bluhm, Robyn. 2005. "From Hierarchy to Network: A Richer View of Evidence for Evidence-Based Medicine." *Perspectives in Biology and Medicine* 48(4):535–47.

Brock, Dan W. 1991. "The Ideal of Shared Decision Making Between Physicians and Patients." *Kennedy Institute of Ethics Journal* 1(1):28–47.

CAST Investigators. 1989. "Preliminary Report: Effect of Encainide and Flecainide on Mortality in a Randomized Trial of Arrhythmia Suppression after Myocardial Infarction." *New England Journal of Medicine* 321(6):406–12.

Charles, Cathy, Amiram Gafni, and Tim Whelan. 1997. "Shared Decision-Making in the Medical Encounter: What Does it Mean? (Or it Takes at Least Two to Tango)." *Social Science and Medicine* 44(5):681–92.

Choudhry, Niteesh K., Robert H. Fletcher, and Stephen B. Soumerai. 2005. "Systematic Review: The Relationship between Clinical Experience and Quality of Health Care." *Annals of Internal Medicine* 142(4):260–73.

Djulbegovic, Benjamin, Gordon H. Guyatt, and Richard E. Ashcroft. 2009. "Epistemologic Inquiries in Evidence-Based Medicine." *Cancer Control* 16(2):158–68.

Emanuel, Ezekiel G., and Linda L. Emanuel. 1992. "Four Models of the Physician-Patient Relationship." *JAMA* 267(16):2221–6.

The Evidence-Based Medicine Working Group. 1992. "Evidence-Based Medicine: A New Approach to Teaching the Practice of Medicine." *JAMA* 268(17):2420–5.

Feinstein, Alvin R., and Ralph I. Horwitz. 1997. "Problems in the 'Evidence' of Evidence-Based Medicine." *American Journal of Medicine* 103(6):529–35.

Fulford, K.W.M., Ed Peile, and Heidi Carroll. 2012. *Essential Values-Based Practice: Clinical Stories Linking Science with People.* Cambridge: Cambridge University Press.

Giacomini, Mita. 2009. "Theory-Based Medicine and the Role of Evidence: Why the Emperor Needs New Clothes, Again." *Perspectives in Biology and Medicine* 52(2):234–51.

Goldenberg, Maya J. 2006. "On Evidence and Evidence-Based Medicine: Lessons from the Philosophy of Science." *Social Science and Medicine* 62(11):2621–32.

Henry, Stephen G. 2006. "Recognizing Tacit Knowledge in Medical Epistemology." *Theoretical Medicine and Bioethics* 27(3):187–213.

Howick, Jeremy. 2011. *The Philosophy of Evidence-based Medicine.* Oxford: Wiley-Blackwell.

Jonsen, Albert, Mark Siegler, and William Winslade. 2002. *Clinical Ethics.* 5th ed. New York: McGraw-Hill.

Jonsen, Albert R., and Stephen Toulmin. 1998. *The Abuse of Casuistry.* Berkeley: University of California Press.

Kempainen, Robert R., Mary B. Migeon, and Fredric M. Wolf. 2003. "Understanding Our Mistakes: A Primer on Errors in Clinical Reasoning." *Medical Teacher* 25(2):177–81.

Lexchin, Joel, Lisa A. Bero, Benjamine Djulbegovic, and Otavio Clark. 2003. "Pharmaceutical Industry Sponsorship and Research Outcome and Quality: Systematic Review." *BMJ* 326:1167–70.

Malterud, Kirsti. 1995. "The Legitimacy of Clinical Knowledge: Toward a Medical Epistemology Embracing the Art of Medicine." *Theoretical Medicine* 16:183–98.

Moumjid, Nora, Amiram Gafni, Alain Brémond, and Marie-Odile Carrere. 2007. "Shared Decision Making in the Medical Encounter: Are We Talking about the Same Thing?" *Medical Decision Making* 27(5):539–46.

Naylor, C.D. 1995. "Grey Zones of Clinical Practice: Some Limits to Evidence-Based Medicine." *Lancet* 345(8953):840–2.

Norman, Geoff, Meredith Young, and Lee Brooks. 2007. "Non-Analytical Models of Clinical Reasoning: The Role of Experience." *Medical Education* 41:1140–45.

Pearce, Neil. 1996. "Traditional Epidemiology, Modern Epidemiology, and Public Health." *American Journal of Public Health* 86(5):678–83.

Pellegrino, Edmund D., and David Thomasma. 1988. *For the Patient's Good: The Restoration of Beneficence in Health Care.* New York: Oxford University Press.

Prendergast, Thomas J. 1995. "Futility and the Common Cold: How Requests for Antibiotics Can Illuminate Care at the End of Life." *Chest* 107(3):836–44.

Rubenfeld, G.D. 2001. "Understanding Why We Agree on the Evidence but Disagree on the Medicine." *Respiratory Care* 46(12):1442–49.

Sackett, David L., William M.C. Rosenberg, J.A. Muir Gray, R. Bryan Haynes, and W. Scott Richardson. 1996. "Evidence Based Medicine: What It Is and What It Isn't." *BMJ* 312:71–2.

Selby Joe V., Anne C. Beal, and Lori Frank. 2012. "The Patient-Centered Outcomes Research Institute (PCORI) National Priorities for Research and Initial Research Agenda." *JAMA* 307(15):1583–84.

Tanenbaum Sandra J. 1993. "What Physicians Know." *New England Journal of Medicine* 329(17):1268–71.

Tinetti, Mary E., and Terri Fried. 2004. "The End of the Disease Era." *The American Journal of Medicine* 116:179–85.

Tonelli, Mark R. 1998. "The Philosophical Limits of Evidence-Based Medicine." *Academic Medicine* 73:234–40.

Tonelli, Mark R. 1999. "In Defense of Expert Opinion." *Academic Medicine* 74(11):1187–92.

Tonelli, Mark R. 2006. "Integrating Evidence into Clinical Practice: An Alternative to Evidence-Based Approaches." *Journal of Evaluation in Clinical Practice* 12(3):248–56.

Tonelli, Mark R. 2010. "The Challenge of Evidence in Clinical Practice." *Journal of Evaluation in Clinical Practice* 16:384–9.

Tonelli, Mark R. 2012. "Compellingness: Assessing the Practical Relevance of Clinical Research Results." *Journal of Evaluation in Clinical Practice* 18:962–7.

Toulmin, Stephen. 2003. *The Uses of Argument.* Cambridge: Cambridge University Press.

Upshur, Ross. 2009. "Making the Grade: Assuring Trustworthiness in Evidence." *Perspectives in Biology and Medicine* 52(2):264–75.

Chapter 2

Evidence-based Medicine versus Expertise

Knowledge, Skills, and Epistemic Actions

Sophie van Baalen and Mieke Boon

Since its inception in the early 1990s, evidence-based medicine (EBM) has been promoted as a way to make clinical practice more scientific, though its epistemology has been criticized on many grounds, for example, for being based on a narrow view of science, which focuses on quantitative, clinical evidence and rule-following instead of basic science, theories, and judgments (Loughlin 2008, 2009; Tonelli 1998; Worrall 2002; Wyer and Silva 2009). This chapter will focus on another line of critique, to wit, EBM's disregard of "expert opinion," in particular the role of physician's expertise in clinical decision-making.

Very shortly after EBM was announced as a "new paradigm for medicine" (Evidence-based Medicine Working Group 1992), Sandra Tanenbaum pointed out some of its philosophical challenges, among which is a misrepresentation of clinical reasoning: "In an act of *interpretation*, not *application*, physicians make clinical sense of a case, rather than placing it in a general category of cases. As interpreters, physicians draw on all their knowledge, including their own experience of patients and laboratory-science models of cause and effect" (Tanenbaum 1993, p. 1269, our emphasis). She argues that clinical medicine inherently has to deal with the uncertainty of a situation and with incomplete information. For physicians to make wise decisions requires them to necessarily rely on "personal knowledge," including experience and sensory input.

In a similar fashion, Mark Tonelli (1998) defended "expert opinion" by first arguing that expertise is wrongfully put on the lowest rung in the "hierarchy of evidence" ladder because it is different *in kind* rather than *in degree* from the other types of evidence that EBM ranks. According to him, expert opinion is crucial to overcome the *epistemic* gap between the outcomes of population-based clinical trials and those of the individual patient, as well as the *normative* gap between what is regarded as the "best" treatment according to EBM (i.e., the treatment that gives the best outcome to a group of patients with a particular disease in a clinical trial) and what will be the "best" treatment for an individual patient. With a clinical case description, Tonelli illustrates how an experienced physician uses "all relevant kinds of medical knowledge, along with patient goals, values and preferences, in order to reach the best possible decision for the patient-at-hand" (Tonelli 2006, p. 73). The role of expertise in such decision-making requires a physician to be aware of the strengths and weaknesses of each kind of medical knowledge and to cope with the fact that evidence derived from clinical research cannot be prescriptive but still requires experience and pathophysiological knowledge.

One of EBM's initial aims was to improve the quality of clinical decision-making by minimizing bias, subjectivity, and uncertainly through statistical evidence and rule-based reasoning (Evidence-based Medicine Working Group 1992). As Tanenbaum (1993) and Tonelli (1998, 2006) argue, EBM falls short in its initial aims when neglecting the crucial role of expert opinion in making clinical decisions about individual patients. Together with other critiques, this has led proponents of EBM to ask whether it is "a movement in crisis" and to contend that "evidence based medicine has not resolved the problems it set out to address (especially evidence biases and the hidden hand of vested interests)" (Greenhalgh, Howick, and Maskrey 2014, p. 5). Trisha Greenhalgh et al. (2014) argue that in order for EBM to overcome this crisis, it needs to refocus its attention on training doctors to use intuitive reasoning based on experience, followed by formal EBM methods to "check, explain, and communicate diagnoses and decisions" and to share uncertainty with patients. Furthermore, they argue that drawing on a "wider range of underpinning disciplines," including qualitative research, would enrich the field.

In other words, EBM proponents admit that EBM falls short as a guiding principle for clinical decision-making. Tonelli, in turn, admits that the danger of overly relying on the authority of medical experts is that experts may become "authoritarian," rigid, and not receptive to new evidence and insights. In addition, expert opinion does not necessarily provide good grounds for decision-making, as experts can disagree or have biased opinions. Thus, opponents and proponents of EBM are showing signs of rapprochement in the sense that both sides acknowledge that evidence from clinical research alone is not sufficient to make diagnosis and treatment decisions for individual patients, whereas reliance on expert opinion holds the danger of an unwanted return to "authority-based" decision-making. The challenge that is raised by this rapprochement is to develop an epistemology of clinical decision-making that acknowledges the central role of medical expertise in decision-making, but at the same time allows us to assess the quality of knowledge and reasoning in these practices.

In view of this dilemma, the aim of this chapter is to explicate what aspects of medical expertise allow doctors to make sound clinical decisions. It will be argued that medical expertise does not consist only of formal knowledge, practical skills, or experience (or any combination of these traits), but should also include *cognitive skills*. In other words, medical expertise also involves the ability to perform the *epistemic activities* needed to produce adequate knowledge about individual patients, by combining knowledge and experience from different sources fitted to the specific situation of that patient. Therefore, becoming an expert involves learning to *use* scientific knowledge and medical evidence in clinical decision-making. This is not, however, simply something that is tacitly learned by being immersed in the tradition and authority of existing medical practice. Instead, expertise means that clinicians can actually justify and thus be held accountable for their decisions. This requires, as we will argue, that experts can justify *how* they reach their decisions, in other words, for the underlying reasoning processes. Yet, this idea is seemingly at odds with authors who, when defending the importance of expertise in clinical decision-making, emphasize the role of *tacit knowledge* in these reasoning processes (Henry 2010; Malterud 2001). Our point will be that tacit knowing and reasoning in these accounts are regarded as inarticulate and therefore inaccessible. As a consequence, clinical decision-making remains a mysterious and vague process that cannot be reflected on, running into the danger that the original problem of tradition and authority-based expertise will persist. In order to address this problem, we will analyze the notion of tacit knowledge and its role in expertise by starting from Collins and Evans's (2007) skill-based account of expertise. Next, in order to open up and evaluate the tacit dimension in their notion of expertise, we will revisit Polanyi's (1958) original account of tacit knowing. As we will show, Polanyi's more subtle notion allows for better understanding how clinicians can indeed be held accountable for their decisions. As well as an epistemological dimension, "being held accountable for decision-making in diagnosis and treatment" involves an ethical dimension. Therefore, we have dubbed this type of accountability the *epistemological responsibility* of clinicians (van Baalen and Boon 2015), which we consider an inherent aspect of an enriched understanding of medical expertise. Based on our analysis of the role of tacit knowledge in expertise, we are in a position to explain in more depth the meaning of epistemological responsibility of clinicians that aims to overcome the dangers of tradition and authority-based notion of expertise. Finally, we believe that the proposed account of expertise in terms of epistemological responsibility is relevant for two reasons: first, in understanding how the quality of clinical decision-making can be accounted for and, second, in specifying learning-aims of the education of medical students.

EPISTEMOLOGICAL RESPONSIBILITY
IN CLINICAL PRACTICE

Recently, we have argued that medical doctors have a professional responsibility to come up with the best possible diagnosis and treatment plan for individual patients, by generating knowledge about the disease of a particular patient (van Baalen and Boon 2015). To meet this responsibility, the key epistemological challenge of doctors involves gathering and integrating all relevant yet heterogeneous sources of information, including not just scientific-medical knowledge of diseases and treatments, and diagnostic data on the patient, but also contextual information (e.g., the patient's situation, goals and values, the availability of specific medical equipment and expertise, and the constraints of the medical system) so as to construct a coherent "picture" of the patient's disease and its possible treatments. In addition, this "picture" must be constructed so that it can be used as an *epistemic tool* for reasoning in clinical decision-making. The notion of "epistemic tools" for reasoning about a target-system was coined by (Boon and Knuuttila 2009; Knuuttila and Boon 2011). They propose the notion of scientific models as epistemic *tools* as an alternative to a generally held *representational* notion of scientific models. The analogy we wish to draw is between scientific models and the "picture" of a patient. In their notion of scientific models as epistemic tools, Knuuttila and Boon claim that the process of constructing scientific models (the modeling) is already part of the *justification* of the model, because constructing a model—or in our case the "picture" of the patient—involves *justification of the information* that is built into this "picture" (e.g., information on the individual patient and more general theoretical and clinical information need to be *relevant* as well as *reliable*) and requires that these heterogeneous bits of information are *coherently tied together*. In the vocabulary of our chapter, one could say that modelers have the expertise—knowledge, experience, and cognitive skills—to build models that are at least partly correct about the specific target-system for which the model is constructed. When translated to the medical context, the scientific model (the "picture") constructed by the modeler (a medical doctor) is about a specific target-system (an individual patient) and functions as an epistemic tool that allows for scientific reasoning about the target-system. Hence, instead of truly representing the target-system, the scientific model is a tool that allows for scientific reasoning about the system. Similarly, instead of finding objective "truth" about the individual patient by scientific approaches such as EBM, "pictures" of individual patients constructed by medical experts allow for reasoning about the patient in clinical decision-making, for instance, in formulating relevant and testable questions and hypotheses that eventually guide toward proper diagnoses and predicting which cure may work as part of the treatment plan, or in explaining why a treatment causes side effects. In clinical practice, the concept "model" has a different connotation; therefore, the word "picture" is used to characterize similar epistemic processes.

We have argued that the epistemological challenge for which doctors bear responsibility is how they build up the "picture" of each patient individually. This epistemic tool should therefore be evaluated in relation to relevant epistemic criteria, such as logical consistency and coherence with relevant knowledge. Another important epistemic criterion is utility for a specific situation. Therefore, contextual and personal information, such as the availability of a certain device in a hospital, a doctor's experience, and a patient's preferences, are all relevant in the epistemic activities of generating and using this "picture" toward diagnosis and treatment, because this information has an impact on how to make the best possible diagnosis and treatment decision for the individual patient in a *specific* situation.

Due to the heterogeneous character of the various aspects playing a role in the construction of an epistemic tool, it is not possible to construct this tool by means of rule-based reasoning only. Instead, constructing an epistemic tool (the "picture" of a patient) requires complex reasoning and assessment of evidence. Therefore, we have proposed that, instead of deferring responsibility to general clinical guidelines, as at least some interpretations of EBM seem to suggest, doctors should consider themselves *epistemologically responsible* for producing good-quality diagnosis and treatment. In other words, by introducing the concept of epistemological responsibility, we allow for a richer and less rigid epistemology of clinical decision-making that leaves space for alternative modes of reasoning better suited to the epistemological challenges of clinical decision-making. However, in spite of the space for personal judgments by medical experts defended here, we wish to avoid a return to the justification of decisions on the basis of cognitively empty authority or "professional opinion." By introducing the notion of epistemological responsibility, we point at specific responsibilities of clinicians toward the best possible execution of *epistemic activities* in diagnosis and treatment and at the accountability of clinicians for the quality of their decisions. Specific skills are required to perform epistemic activities responsibly, and these skills are an important aspect of medical expertise. A further characterization of these skills will therefore help clarify what it means for a clinician to be epistemologically responsible and to help assess the quality of expert decision-making.

Lorraine Code introduces the concept of epistemic responsibility as a "potential new focal point for theory of knowledge" (Code 1984, p. 29). Code draws analogies between ethical and epistemological reasoning processes to show how an epistemological inquiry can be approached by a study of intellectual virtues, instead of searching for foundations as in traditional foundationalism. Important for our argument is Code's insight that, first, cognitive agents (such as doctors) have an important degree of choice when it comes to reasoning, and second, they are accountable for these choices.[1] Therefore, in contrast to passive information-processers that are at best *reliable*, these agents should be evaluated in terms of responsibility. Following Code, we argue that the epistemological tasks of doctors involve a considerable amount of choice, deliberation, and justification, for which they are held responsible in our account of medical decision-making. By introducing these two notions, "epistemological responsibility" and "knowledge of a specific patient as an epistemic tool for reasoning," we have shifted the focus from "epistemic truth" and passive rule-following to "epistemic use" and active knowledge-construction. Within the epistemological responsibility framework of medical decision-making, it makes sense to focus on what doctors actually do when they generate knowledge of a patient and make diagnostic and treatment decisions, in other words to the *epistemic activities* they perform.

In short, we have argued that clinicians are epistemologically responsible for constructing a "picture" of each individual patient from heterogeneous sources of evidence and using that "picture" as an epistemic tool for clinical reasoning. However, expecting doctors to bear responsibility for the execution of epistemic activities requires a detailed account of how competence in these actions is developed, how epistemic activities relate to medical expertise, and how they can be assessed, in order for doctors, students, and policy makers to know what a clinician's accountability involves. Our framework provides useful leads for developing such an account: by shifting the attention from *what is known* to the *knower*, we return to a more doctor-centered account of clinical decision-making, and by focusing on *epistemic tools*, *epistemic use*, and *epistemic activities*, we point out that what matters for clinical decision-making is not only the knowledge and information that is used, but also *what a doctor does* with it, in other words, how that knowledge and information are used in the reasoning process to reach clinical decisions. Performing actions and using tools *well* require skills that are developed as a part of a professional's *expertise*. In explicating how clinical expertise is developed and what it means to develop skills in performing epistemic activities, we will first evaluate Collins and Evans's skill-based account of expertise.

MEDICAL EXPERTISE

Collins and Evans have developed a view of expertise in which expertise is not solely defined by the amount of formal knowledge possessed by individuals, but as something *practical*:[2] "something based in what you can do rather than what you can calculate or learn" (Collins and Evans 2007, p. 23). Central to their thesis on expertise is an immersion within a certain society, which is necessary in order to gain expertise. They distinguish between two levels of specialist expertise, *interactional* and *contributory* expertise. They define interactional expertise as "expertise in the *language* of a specialism, in the absence of expertise in its *practice*" (ibid, p. 28). In other words, only immersion in the world of language is needed to acquire this kind of expertise, which requires only a "minimal body" that fulfills only the requirements that are necessary to learn a language. This is what they call "the minimal embodiment thesis" (ibid, p. 79). In contrast, they define contributory expertise as "enabling those who have acquired it [contributory expertise] to *contribute* to the domain to which the expertise pertains: contributory experts have the ability to *do* things within the domain of expertise" (ibid, p. 24). According to Collins and Evans, this requires full embodiment[3] and immersion in a social group of experts of a domain.

Collins and Evans point out that all specialist expertise (interactional as well as contributory) involves a great deal of *specialist tacit knowledge*,[4] defined by them as "things you know how to do without being able to explain the rules for how you do them" (ibid, p. 13). Therefore, in order to acquire any degree of specialist expertise requires interactive immersion in a specialist culture or *enculturation* "because it is only through common practice with others that the rules that cannot be written down can come to be understood" (ibid, p. 24). For medical specialists, this would mean that to master medical expertise requires to be immersed in the culture of day-to-day medical practice. An endorsement of this idea is the extensive system of apprenticeship teaching in the education of medical professionals: a large part of what a clinician learns is learned by "doing" in internships, residencies, and fellowships. Collins and Evans present a five-stage model of acquisition of contributory expertise—adopted from Dreyfus and Dreyfus (1986)—in which successive steps (from *novice* to *expert*) represent an increasing internalization of physical skills, exemplified by the process of learning to drive a car (ibid, pp. 24–25). This process hinges on the acquisition and mastering of skills, unselfconsciously recognizing contexts, and unselfconsciously acting accordingly. This unselfconscious decision-making in response to a certain context is what is considered as "tacit knowledge" by Collins and Evans. In addition, Collins and Evans emphasize that to understand expertise, one should focus on what experts *do* instead of what they *know*.

In short, in Collins and Evans's account of what makes somebody an expert in a certain specialism involves—besides being familiar with the epistemic content—skills and enculturation, which in the case of high-level experts are largely tacit. Their account presents important reasons for the crucial role of (medical) expertise, but may unintendedly vindicate nontransparent authority-based reasoning in medical decision-making. Therefore, in our view, it has two shortcomings. First, it does not address acquisition and cultivation of *cognitive skills* required to perform *epistemic actions*, and second, it does not provide an adequate account of tacit knowledge to analyze the quality of medical reasoning.

In everyday clinical practice, doctors are continuously performing epistemic actions, for example, hypothesizing which treatments will be beneficial, predicting the risk of adverse effects, or adjusting treatment plans and interpreting new information in light of what is already known about a patient. Collins and Evans focus mainly on the acquisition of physical skills and on enculturation (i.e., understanding and use of language and unwritten social rules) into the relevant group practices. Although we agree with them that in medical practice skills (such as surgical skills and communication skills), epistemic content (such as basic biomedical knowledge, knowledge of treatments, and up-to-date knowledge of clinical science) and acquaintance with the medical (hospital) culture (e.g., traditions, hierarchy, and behavioral etiquette) are all highly important, our point is that focusing on these aspect of expertise leaves out another aspect that is particularly crucial for clinical decision-making, which is the ability to perform specific epistemic actions related to medical reasoning at an expert level. Epistemic actions are similar to physical actions in that they generate a transformation from the initial state to another—for epistemic actions this concerns a change in an agent's mental state, whereas a physical action results in the change of an actor's environment. Furthermore, both epistemic and physical actions are temporarily confined and are directed toward achieving some goal (Neth and Muller 2008).[5] Just as expertise in a certain specialism involves a specific set of physical actions that a person should master, it also involves a set of epistemic actions that experts are able to perform skillfully. Consequently, Collins and Evans's ideas about the acquisition of expertise through experience and apprenticeships should be extended to include the acquisition and deployment of *cognitive skills.*

The second shortcoming of Collins and Evans's account of expertise is their notion of tacit knowledge. They assign a pivotal role to tacit knowledge in expertise, for example, by describing expertise as an ability to tacitly make decisions in response to a certain situation or context. However, in their definition, tacit knowledge is inarticulate and therefore inaccessible for inquiry and evaluation. As a consequence, their account obscures an analysis of what it involves to make these decisions and what it means to be an expert in that respect. The problem with thinking about the skillful execution of epistemic actions in this way seems to be that expert decision-making and the knowledge used in the process become increasingly inaccessible to others, which increases the likelihood of referring to authority rather than epistemic quality in clinical decision-making.

In contrast, a fruitful interpretation of "tacit knowledge" in clinical reasoning for diagnosis and treatment would focus on *what kind* of knowledge is used and *how* reasoning is performed tacitly, so that its strength, relevance, and application can be evaluated. For example, Michael Loughlin (2010) argues that "features of our knowledge that function tacitly in many contexts can, without contradiction, be made the object of explicit attention in others" (Loughlin 2010, p. 298). Loughlin argues that it is a mistake to assume that tacit knowledge is completely inarticulate in all situations. Stephen Henry (2010), analyzing how a physician in a neurological exam focuses on observing neurological symptoms while being tacitly aware of a patient's body part, writes: "Whether information is tacit or explicit has less to do with its content than it does with how it functions in a particular situation" (Henry 2010, p. 294). In other words, tacit knowledge is not essentially different from explicit knowledge, and therefore, knowledge that functions tacitly in one situation can be reconstructed in another situation to study how it informs clinical decision-making. Hence, an account of tacit knowledge in medical reasoning need not be implicit and mysterious.

Summing up, Collins and Evans show that tacit knowledge plays a pivotal role in expertise, but their view of tacit knowledge as inarticulate does not clarify how to evaluate the reasoning process. Referring to inarticulate tacit knowledge as a justification of clinical decisions and actions is not enough to hold clinicians accountable for the decisions they make. In order to ensure the quality of clinical decision-making for diagnosis and treatment, a better understanding of tacit knowledge is needed, which is why we will revisit the original definition by Michael Polanyi.

TACIT KNOWLEDGE

Michael Polanyi first introduced the concept "tacit knowledge" in his book *Personal Knowledge: Towards a Post-Critical Philosophy* (1958) and further explicates this notion in *The Tacit Dimension* (1966). His idea has been widely taken up to make sense of situations in which experience, know-how, and practice play important roles. For example, the notion of tacit knowledge has been important in the (sociological) analysis of scientific practices, such as gravitational physics labs, and the realization that in such practices, in addition to knowledge, theories, and concepts, skills and personal contacts are crucial (for example, see Collins 2001). However, over the years, tacit knowledge has come to mean "inarticulate," and therefore as not transferrable and inaccessible by others. This conception of tacit knowledge leads to obfuscating and mystifying expertise and is therefore seemingly unsuitable as a concept to clarify expert knowing. In contrast, Polanyi's original notion is much more detailed and refined and therefore appears to be very helpful in better understanding expertise.

In Polanyi's epistemology, knowing comprises two types of awareness: the *subsidiary* and the *focal* (Polanyi 1958, p. 57; also see Mitchell 2006, p. 71). Focal awareness is the conscious object of our awareness, whereas subsidiary awareness includes a variety of background clues that enable focal awareness. An example to understand these two kinds of awareness is the recognition of somebody's face: "We know a person's face, and can recognize it among a thousand .... Yet we usually cannot tell how we recognize a face we know" (Polanyi 1966, p. 4). Polanyi goes on to explain that faces consist of a collection of particular physiognomics such as a nose, a mouth, eyes, ears, etc. In recognizing a face, we are not consciously aware of all the particulars of a face, but rather by integrating our awareness of its particulars, as subsidiaries, in order to recognize the whole, the face that is our focal awareness. This is, according to Polanyi, "the outcome of an active shaping of experience performed in the pursuit of knowledge" (ibid, p. 6). Therefore, in tacit knowing, the *knower* is the third essential element. These three elements make up a *from-to* relationship: the knower "attend[s] *from* the subsidiaries *to* the focal target" (Mitchell 2006, p. 73).

According to Polanyi, the *use of tools* (e.g., technological instruments such as a probe) also involves "incorporation" in our subsidiary awareness. He calls this process, in which the use of a tool moves from focal to subsidiary awareness, "indwelling." For example, when an ultrasound is made, a doctor dwells in the probe to observe organs that remain otherwise unseen: "It is not by looking at things (particulars), but by dwelling in them that we understand their joint meaning" (Polanyi 1966, p. 20). As we have characterized the picture doctors develop of their patients as *epistemic tools*, we can understand the use of these tools in a similar fashion. The use of epistemic tools or theories become part of our subsidiary awareness; it is "interiorized" to understand something, extending the cognitive apparatus of the knower. It is, however, important to note that subsidiary awareness is *not necessarily* unconscious awareness, because the use of the tool, such as the ultrasound apparatus or the epistemic tool, can always be brought back to the focal awareness, for instance when inspecting its proper functioning or adequateness. But this is at the cost of focal awareness of the observed object. When one draws their attention away from the focal attention to the subsidiaries, the conception of the entity—which is recognized or understood through the subsidiary use of technical or epistemic tools—is destroyed, in the same way as a pianist who, when turning her attention from the music to the separate notes and the movement of her fingers, will likely become confused and cease to play (ibid, p. 73).

In other words, part of the "triad" will remain unspecifiable, as it is impossible to focus one's direct attention on it. In summary, Polanyi recognizes that although during the act of knowing, it is impossible to articulate which particulars make up the background clues of the subsidiary awareness, it is very well possible to make a reconstruction (although this is usually not a complete account of how knowledge is perceived). Furthermore, Polanyi explicates the relationship between what is tacitly known and what is consciously known.

We will now apply Polanyi's original conception of tacit knowing to analyze skillful execution of epistemic actions in clinical decision-making. Conversely, this analysis serves as an illustration of Polanyi's understanding of tacit knowledge. Within our framework of epistemological responsibility in clinical decision-making, epistemic actions can be divided into three broad activities. First, the gathering and critical assessment of relevant information, second the *construction* of a coherent "picture" of the individual patient from these heterogeneous pieces of information, and third the application and adaptation of this "picture" (van Baalen and Boon 2015). Being competent in these specific epistemic activities is an important aspect of the expertise of doctors, which allows them to face the main epistemological challenges of clinical practice responsibly.

Gathering and Assessment of Information

Physicians need to gather relevant information by searching the literature and textbooks, by ordering lab tests and images, by physical examination of the patient, and by taking the patient's medical history. In all of these situations, the *relevance* and *utility* of the information for the particular patient needs to be assessed—for example, in the case of a literature search. As the clinical encounter usually commences with the question "what brings you here today?" (Montgomery 2006), the first set of information that doctors collect is a patient's account of their symptoms. While searching the scientific literature and textbooks, the particular signs and symptoms of a patient function as tacit knowledge, residing in a clinicians' *subsidiary awareness*, from which clinicians are able to direct their *focal awareness*, to the question to be answered, thereby assessing the relevance of the information that is coming across. In short, deciding on what information to use in a particular case, which is an *epistemic action*, relies on the tacit use of knowledge of particular aspects of a patient. Conversely, when doctors attend to their patients to gather information through physical exam and medical history taking, theoretical information functions as *subsidiary awareness*, as background information about possible diagnoses and pathophysiological mechanism that cause the symptoms, from which clinicians direct their *focal awareness* to the conversation with the patient and the proceedings of the exam. As such, the subsidiary information functions as a conceptual and theoretical framework that allows coming up with and asking the right questions, interpreting the answers, and subsequently following up on them. Thus, in contrast to the appraisal of textual information, the gathering of particular patient information requires the subsidiary use of theoretical knowledge while directing one's focal awareness to performing epistemic activities such as prioritizing certain questions and interpreting answers in light of prior knowledge.

As Collins and Evans contend, the subsidiary information that experts use tacitly is often selected unselfconsciously. However, it is a mistake to conclude, as they seem to suggest, that subsidiary information cannot be identified and evaluated and that choices that were made unselfconsciously are unjustifiable. It is possible to check, for instance, whether the theoretical information used in clinical decisions was up-to-date and relevant for the specific case at hand. Examples of such inquiries are: Is the information based on good-quality science, and if such scientific support is not available, what else warrants the use of that piece of information? When the patient's story is subsidiary, which information is given priority and for what reason? Did the physician obtain the story firsthand? Is the patient considered to be a trustworthy source? Did the patient provide enough information? These examples illustrate that, although knowledge is used tacitly and involves a considerable amount of choice and variability, physicians can be held accountable for the quality of knowledge they use and how they use it.

Constructing the Epistemic Tool in the Sense of a "Picture" of the Patient

In order to construct a useful epistemic tool (i.e., the "picture" of an individual patient), it is necessary that the information gathered by a clinician—the written, scientific, or general knowledge as well as patient-specific evidence such as lab results, imaging, and clinical history—is comprehensive, relevant, and up-to-date. Subsequently, the construction and use of such an epistemic tool involves tacit knowledge and epistemic activities. For example, in fitting the different pieces of information together, these pieces need to be weighted and mutually adjusted, which requires epistemological choices and an active involvement of the epistemic agent. Tacitly, thus in their subsidiary awareness, prior experience with similar patients is used to draw analogies with the current patient. Through the subsidiary awareness of biomedical knowledge, the relevance of certain signs or symptoms can be assessed and weighted in relation to others. By subsidiary awareness of results from earlier tests, the results of new tests can be interpreted and fitted in with what is already known. When a disease develops over time, an up-to-date picture of a patient is obtained through subsidiary awareness of the prior instances of the disease in that patient. Such uses of tacit knowledge allow doctors to fit together relevant information (the "particulars") according to epistemic criteria, such as internal logical consistency, coherence with background knowledge, and comprehensiveness of information, thus directing their focal awareness to "whole," the resulting "picture" that is under construction.

The Adaptation and Application of a "Picture" of the Patient

The resulting "picture" functions as an epistemic tool that allows epistemic actions, such as further reasoning, hypothesizing, and making diagnostic and treatment decisions.[6] In these cases, the epistemic tool, similar to physical tools, is used tacitly, focusing the attention on the *goal* of performing epistemic action (e.g., to draw up a possible diagnosis). In Polanyi's words, talking about physical tools in quite similar ways as Collins and Evans, "We are attending to the meaning of its impact on our hands in terms of its [the physical tool's] effect on the things to which we are applying it" (Polanyi 1966, p. 13). The use of such a tool requires *skills*. Mastering cognitive skills in clinical practice is in that sense on the same footing as physical skills, as also recognized by Polanyi: "The art of the expert diagnostician ... we may put it in the same class as the performance of skills, whether artistic, athletic or technical" (Polanyi 1966, p. 6). In other words, the cognitive skills that enable a medical expert to perform diagnosis are similar to the skills of painters handling their brush and paint, a football player skillfully passing his opponent with a ball, or a surgeon closing a wound with sutures. And, similar to other skills, different levels of mastery can be obtained, from novice to expert level, through practice and experience. Analogous to the responsibility of clinicians to ensure their competence in physical skills,[7] doctors have a responsibility to develop and cultivate their cognitive skills, which is part of their *epistemological responsibility*.

Finally, Polanyi's notion of tacit knowledge enables us to better understand two other important aspects of medical expertise playing a role in the three epistemic activities just listed, which set experts apart from novices. First, the ability of experts to deploy large amounts of information, including textbook knowledge and experience in their subsidiary awareness while focusing their attention on the patient at hand. When faced with cognitive tasks, a range of earlier acquired knowledge and information is tacitly invoked through which experts focus on the epistemic action. In contrast, novices still need to direct their focal attention to particulars such as theoretical information while talking to the patient or focus on patient particularities while searching for information. Analogous to the piano player who becomes confused when focusing on separate notes or finger movements, this focusing of attention impedes the execution of epistemic actions. Thus, similar to the mastery of *physical skills*, doctors and other experts master cognitive skills through practicing and improving their epistemic actions, being able to include more and more knowledge tacitly while being focused on the specific action.

A second aspect that sets experts apart from novices is the tacit recognition of relevant patterns. By having gained experience from diagnosing and treating many patients, expert clinicians have collected a wide range of exemplary cases including signs and symptoms, the diagnosis that followed, the choice of treatment, and the clinical outcome. Clinicians are able to group these cases together according to similarities and differences between cases and distill what combination of signs and symptoms, measurements, and other evidence suggests that a case belongs to a certain group. In interaction with patients or evidence, these patterns enable to recognize possible diseases and direct courses of action. Tacit pattern recognition can be compared to chess grandmasters—an example used by Polanyi—who recognize patterns in the way the pieces are positioned on the board without having to assess the position and possible course of each individual piece and who can act accordingly. Such a catalogue of patterns is build up through practice and from prior experience. Important to our argument is that, very similar to theoretical knowledge acquired in formal education, this catalogue of patterns operates as subsidiary awareness, allowing physicians (and chess grandmasters) to recognize what is going on and how to handle it. This is not to say that pattern recognition is a justification for a medical decision at the same level as theoretical and evidence-based knowledge. Instead, it is an indispensable mechanism that fills the gaps not yet covered by formally tested knowledge. Furthermore, pattern recognition enables physicians to direct their reasoning and search for further information. Although patterns are recognized tacitly, instantly, and unselfconsciously, they are open to evaluation, and physicians are responsible for the conclusions they reach on the basis of a recognized pattern.

EPISTEMOLOGICAL RESPONSIBILITY OF EXPERTS IN CLINICAL DECISION-MAKING

One of the main criticisms against the epistemology of EBM is that it wrongly assumes that the role of expertise in medical decision-making can and should be reduced in order to increase objectivity and quality. Nevertheless, EBM has rightfully challenged "authority-based" decision-making, by demanding quality in accordance with scientific standards. In this chapter, we have reconciled EBM's call for quality decision-making with medical expertise. We have argued that what seems to be missing in current accounts of medical expertise is an understanding of the role of the reasoning process in medical decision-making about individual patients. Therefore, we have argued that a more detailed analysis of what medical expertise entails is needed. We have first turned to Collins and Evans who have highlighted that expertise is about what experts can do rather than primarily about what they know and who have pointed out that tacit knowledge is a central trait of experts. In addition to their analysis, which focuses on cultural integration, mastering physical skills, and epistemic content, we have argued that an understanding of expertise must also include the role of cognitive skills. Furthermore, it should include an explanation of *how* tacit knowledge allows experts to practice their expertise. Michael Polanyi's original notion of tacit knowledge is richer and more detailed than that is usually put forward by authors defending medical expertise. Polanyi describes tacit knowing as a triad consisting of three components: the knower, their subsidiary awareness, and their focal awareness. This triadic conception allows for a more fine-grained analysis of the epistemic actions that make up decision-making for diagnosis and treatment, and of how tacit knowledge is a specific trait of experts in comparison to novices.

An important result of elucidating the role of tacit knowledge in expertise is that it clarifies how medical experts can be held accountable for epistemic activities in medical decision-making. We have argued that cognitive skills and competence in epistemic actions are considered to be crucial aspects of medical expertise and doctors have a responsibility to develop, acquire, and cultivate these skills and competence in a similar way as they have a responsibility to develop, acquire, and cultivate physical skills such as surgical skills. We follow Collins and Evans in their claim that tacit knowledge plays a pivotal role in expertise, but we refuse to regard it as inarticulate and therefore hidden from evaluation. Rather, by referring to Polanyi's original conception of tacit knowing, we showed that there is a distinction between subsidiary and focal attention and that knowledge in one's subsidiary awareness can be opened up to scrutiny, so that doctors can be held accountable for what information is used tacitly, be it textbook knowledge, outcomes from RCTs, or experience. What information is used, in either one's subsidiary or focal awareness, should be justifiable by the knower. By giving an account of how medical experts employ tacit knowledge and how they become skilled in performing epistemic activities, we have detailed what the epistemological responsibility of doctors entails.

In conclusion, a crucial aspect of medical expertise is competence in performing epistemic activities, which requires cognitive skills, and similar to physical skills, mastering cognitive skills entails internalization, or tacit knowledge. Doctors can and should be held accountable for how well they perform epistemic activities in decision-making. This means that they can be held responsible for acquiring, developing, and maintaining competence in cognitive skills, and also that they bear responsibility for the quality of the knowledge and reasoning processes. This is what we have called their "epistemological responsibility."

NOTES

1. Code argues that this epistemology would parallel an ethics based on moral virtues, and hence investigates which *intellectual virtues* an epistemologically responsible agent should possess. In order to connect to theories of expertise, we will focus on skills rather than virtues, but still similar to Code, emphasize the importance of *cognitive skills*.

2. Collins and Evans introduce this "wisdom-based or competence-based model" as an alternative to the way that expertise has been understood, which has developed over the last half-century, inspired by phenomenological philosophers such as Heidegger and Merleau-Ponty, and ideas from philosophers like Polanyi and Wittgenstein.

3. One thing that might be argued for medical expertise is that only a minimal body is required for epistemic action, and therefore, people with only *interactional medical expertise* are capable of performing clinical judgment. However, the point that we want to make in this chapter is not that medical expertise is a particular kind of expertise that can be fully characterized by considering the relevant epistemic activities. Rather, we argue that, in order to fully explicate medical expertise, our account of expertise should be widened to include, besides knowledge and physical skills, also cognitive skills.

4. Collins and Evans contrast "specialist expertise" to "ubiquitous expertise," the latter including "all the endlessly indescribable skills it takes to live in a human society" (Collins and Evans 2007, p. 16). They argue that ubiquitous expertise requires *ubiquitous* tacit knowledge, which is acquired through immersion in society in general, whereas specialist tacit knowledge requires immersion in a community of specialists.

5. Neth and Muller distinguish between *practical* versus *theoretical* action because earlier work in cognitive science used *epistemic actions* to describe "physical actions that improve cognition by facilitating or reducing the need for internal computations" (Neth and Muller 2008, p. 994). We adopt the term "epistemic action" to emphasize the relation with the two other notions introduced to account for the cognitive side of "expertise," *epistemic tools* and *epistemological responsibility*.

6. It is important to note that an epistemic tool is a "tool in its own making" (Knuutila and Boon 2011, p. 318): as it is being used to formulate questions and hypotheses, it is adapted to newly gained insights. In other words, it *develops* through its use, in an interaction between tacitly held knowledge, experience, explicit reasoning, in consultation with others, and by including new information. In addition, the epistemic criteria mentioned above (consistency, coherence, specificity, and comprehensiveness) remain continuously relevant, in both the construction and use of an epistemic tool. Thus, although the epistemic actions differ between construction and use of an epistemic tool, a distinction between the two are analytical rather than being two consecutive, distinct steps in the diagnostic process.

7. In the Dutch health care law, doctors carry a personal responsibility to develop and cultivate their competence in medical procedures ("geneeskundige handelingen") that are reserved to be solely performed by professional clinicians.

REFERENCES

Boon, Mieke, and Tarja Knuuttila. 2009. "Models as Epistemic Tools in Engineering Sciences: A Pragmatic Approach." In *Philosophy of Technology and Engineering Sciences*, edited by Anthonie Meijers, 687–720. North Holland: Elsevier Science & Technology.

Code, Lorraine. 1984. "Toward a Responsibilist Epistemology." *Philosophy and Phenomenlogical Research* 45:29–50.

Collins, Harry, and Evans, Robert. 2007. *Rethinking Expertise.* Chicago, London: The University of Chicago Press.

Collins, Harry M. 2001. "Tacit Knowledge, Trust and the Q of Sapphire." *Social Studies of Science* 31(1):71–85.

The Evidence-Based Medicine Working Group. 1992. "Evidence-Based Medicine: A New Approach to Teaching the Practice of Medicine." *JAMA* 268(17):2420–2425.

Greenhalgh, Trisha, Jeremy Howick, and Neal Maskrey. 2014. "Evidence Based Medicine: A Movement in Crisis?" *BMJ* 348. doi: 10.1136/bmj.g3725.

Knuuttila, Tarja, and Mieke Boon. 2011. "How Do Models Give Us Knowledge? The Case of Carnot's Ideal Heat Engine." *European Journal for Philosophy of Science* 1(3):309–334.

Loughlin, Michael. 2008. "Reason, Reality and Objectivity–Shared Dogmas and Distortions in the Way Both 'Scientistic' and 'Postmodern' Commentators Frame the EBM Debate." *Journal of Evaluation in Clinical Practice* 14(5):665–671.

Loughlin, Michael. 2009. "The Basis of Medical Knowledge: Judgement, Objectivity and the History of Ideas." *Journal of Evaluation in Clinical Practice* 15(6): 935–940.

Loughlin, Michael. 2010. "Epistemology, Biology and Mysticism: Comments on 'Polanyi's Tacit Knowledge and the Relevance of Epistemology to Clinical Medicine'." *Journal of Evaluation in Clinical Practice* 16(2):298–300.

Mitchell, Mark T. 2006. *Michael Polanyi: The Art of Knowing.* Wilmington, Delaware: ISI books.

Montgomery, Katherine. 2006. *How Doctors Think: Clinical Judgment and the Practice of Medicine.* New York: Oxford University Press.

Neth, Hansjörg, and Thomas Muller. 2008. "Thinking by Doing and Doing by Thinking: A Taxonomy of Actions." *Paper presented at the 30th Annual Meeting of the Cognitive Science Society*, Austin, TX.

Polanyi, Michael. 1958. *Personal Knowledge: Towards a Post-Critical Philosophy* (1974 ed.). Chicago and London: The University of Chicago Press.

Polanyi, Michael. 1966. *The Tacit Dimension.* Chicago and London: The University of Chicago Press.

Tanenbaum Sandra J. 1993. "What Physicians Know." *New England Journal of Medicine* 329(17):1268–1271.

Tonelli, Mark R. 1998. "The Philosophical Limits of Evidence-Based Medicine." *Academic Medicine* 73:234–240.

Tonelli, Mark R. 2006. "Evidence Based Medicine and Clinical Expertise." *JAMA* 8(2):71–74.

van Baalen, Sophie, and Mieke Boon. 2015. "An Epistemological Shift: From Evidence-Based Medicine to Epistemological Responsibility." *Journal of Evaluation in Clinical Practice* 21(3):433–439.

Worrall, John. 2002. "What Evidence in Evidence-Based Medicine?" *Philosophy of Science* 69(S3):S316–S330.

Wyer, Peter C., and Suzana A. Silva. 2009. "Where is the Wisdom? I–A Conceptual History of Evidence-Based Medicine." *Journal of Evaluation in Clinical Practice* 15(6):891–898.

Chapter 3

Phenomenology, Typification, and Ideal Types in Psychiatric Diagnosis and Classification

Anthony Vincent Fernandez

It is now well known that contemporary psychiatry finds itself in a state of crisis. In spite of an increase in interrater reliability—that is, the likelihood of two or more clinicians granting the same diagnosis to the same patient— recent editions of the *Diagnostic and Statistical Manual of Mental Disorders* (*DSM*) have failed to provide valid categories of disorder. Receiving a *DSM* diagnosis does not supply one with accurate predictions of course of illness or treatment response. Furthermore, a diagnosis supplies little, if any, information about distinct neurobiological substrates or genetic markers. In light of this, the state of psychiatric classification has been criticized from within the field of psychiatry itself (Insel 2013), and psychiatrists have begun developing alternative systems of classification, such as the National Institute of Mental Health's Research Domain Criteria initiative (Cuthbert and Insel 2013; Cuthbert 2014).

While the rise of this critical stance within mainstream psychiatry is fairly new, phenomenologically trained psychologists, psychiatrists, and philosophers have been criticizing the *DSM* system of classification for the past few decades. They are especially critical of the *DSM*'s polythetic, operational approach, whereby categories of disorder are defined by appealing to sets of easily observable symptoms, and diagnoses are made through structured interviews in which the clinician asks a set of pre-delineated questions with the aim of checking off a set number of these criteria (Schwartz and Wiggins 1986). As these phenomenologists argue, the *DSM* approach suffers from at least three major problems in addition to the failure to adequately validate its categories of disorder. First, there is no attempt to understand the organization among the various criteria included within each category (Parnas and Bovet 2015). Second, the checklist diagnostic method fails to take into account how clinicians intuitively understand and classify their patients, forcing them to engage in a diagnostic approach that conflicts with their everyday modes of understanding (Schwartz and Wiggins 1987b). And third, the categories established in the *DSM* tend to reify, becoming resistant to substantial revision and change (Parnas and Bovet 2015).

Phenomenologists, however, have not limited themselves to criticizing the *DSM* system of classification. They have also proposed alternative models of classification and diagnosis. While not all phenomenologists agree on the best approach, one of the most popular is the ideal type approach developed in the 1980s and 1990s by Michael Schwartz and Osborne Wiggins (1987a). Their approach permeates the literature on phenomenological psychopathology, being appealed to by figures such as Nassir Ghaemi (2003), Matthew Ratcliffe (2015), and Louis Sass and Elizabeth Pienkos (2013), among others. Yet, in spite of their popularity, ideal types are often mischaracterized and misapplied. Their distinctive elements are sometimes ignored, resulting in their conflation with fundamentally different kinds of classificatory systems.

In light of this, I here offer a critical introduction to the role of typification and ideal types in psychiatric diagnosis and classification. In so doing, I take the account developed by Schwartz and Wiggins as my guide, but I also address some of the misleading characterizations that are found in their own work as well as the work of other contemporary phenomenologists.

First, I introduce the notion of typification, clarifying its role in everyday experience and psychiatric practice. Second, I clarify Schwartz and Wiggins's account of ideal types, explaining their role in psychiatric diagnosis and classification. Third, I resolve some of the confusions and misunderstandings that have arisen in the contemporary literature.

THE PHENOMENOLOGY OF TYPIFICATION

In order to adequately articulate the notion of ideal types, we will first need to articulate the process of typification outlined by figures such as Edmund Husserl (1973), Alfred Schutz (1966), and Michael Polanyi (1966; 1974). In Schwartz and Wiggins's approach, the construction and employment of ideal types in psychiatry is founded upon a *scientific* approach to typification, which is itself derived from our *everyday* process of typification.[1] In this section, I outline the basic features of everyday typification, then unpack Schwartz and Wiggins's approach to scientific typification, showing how the latter version plays an important role in psychiatric diagnosis. Having clarified how typifications can be made scientific, I transition to a discussion of ideal types in the following section.

Everyday Typification

On a broadly phenomenological account of human understanding, whenever we experience something *as* something—which is to say, whenever we experience any meaningful entity at all—we necessarily experience it as belonging to a type, kind, or category. In my experience of a table, I immediately experience it *as* a table, rather than as, say, a cluster of shapes and colors that I *infer* is a table. When I walk into a classroom, I immediately experience the person at the front of the room *as* a person, *as* a woman, *as* Latina, *as* a professor, and so on. In most everyday situations, I do not have to weigh the available evidence and puzzle out who someone is. This process of typification is a constitutive feature of human experience and understanding. It gives order to an otherwise chaotic field of awareness, allowing us to discriminate and make sense of the myriad people, objects, events, and situations that we encounter in everyday life.

In order to clarify the notion of typification and provide the groundwork a phenomenological approach to diagnosis, I here outline six of the constitutive features of everyday typification. Typifications are (1) tacit, (2) contextual, (3) anticipatory, (4) sedimented, (5) adaptive, and (6) fuzzy. This list might not be exhaustive, but it highlights the major elements of typification and establishes a groundwork for the following discussions of scientific typifications and ideal types.[2]

First, the typifications that structure my experience are usually tacit. My ability to typify an object as a "chair," for instance, does not imply that I can offer an adequate definition of a chair, or produce a set of criteria by which one could decide whether or not something is or is not a chair. In this sense, everyday typification is largely preconceptual. To say that I employ a typification in my experience of X is not necessarily to say that I have a concept of X. On Schwartz and Wiggins's account, the human capacity to articulate and revise concepts is itself founded upon the everyday capacity for typification. In order to develop a concept, I first need to have some set of objects that show up to me as similar in some respects, and as meaningfully different from other kinds of objects. Only after I have achieved this basic level of discrimination can I begin the more intellectual process of conceptualizing these apparently distinct phenomena.

Furthermore, Schwartz and Wiggins argue that in light of typification's *tacit* functioning, it should be construed as a skill in Polanyi's sense of the term. As he says, "the aim of a skillful performance is achieved by the observance of a set of rules which are not known as such to the person following them" (Polanyi 1974, 49; quoted in Schwartz and Wiggins 1987b, 70). A truly skillful activity—including typification—is usually conducted without any explicit awareness of how the activity is being performed. In many cases, the act of explicit reflection will actually disrupt the flow of the activity.

Second, typifications are contextual. We always typify our environment from a historically embedded, meaningfully situated point of view. The object that I typify as a stockpot while cooking a meal might, in other circumstances, be typified as a container to manage a leaky roof or, when flipped upside down, as a stool for reaching a high shelf. Typifications, in short, are not context independent. They are not objective. What we experience something, or someone, *as*, depends on the particular social context of the encounter, as well as on our own aims and projects. This does not mean that there is no right or wrong way to typify something. Rather, it means that my community and social milieu play a fundamental, normative role in determining the accuracy or inaccuracy of my typification.

Third, typifications are anticipatory. When I typify something, it shows up to me not only as meaningful, but as meaningful in some particular manner. This particular meaningfulness includes a tacit background sense of what the object in question is for, what it is capable of, and what I am supposed to do with it. My typification, therefore, governs the way I approach, interact with, and use the objects that I encounter. Joona Taipale illustrates this in saying, "Even if we have never witnessed *this* water glass breaking into pieces when falling to the floor, we nevertheless expect that *it* would break down if it falls to the ground; even if we have never seen *this* bird spreading its wings and fly, we nevertheless expect *it* to do so if we go close enough" (2015, 151). We do not need to relearn how to interact with each new object. Our expectations about the world are generalized, allowing us to navigate new situations with relative ease.

Fourth, our set of available typifications is sedimented, passed down through generations. Our past experiences and sociohistorical milieu structure our lived world, providing us with taken for granted ways of understanding, engaging with, and even perceiving our environment. This personal and social history lives on by shaping our present and future experience. As Merleau-Ponty says, "This word 'sedimentation' must not trick us: this contracted knowledge is not an inert mass at the foundation of our consciousness" (Merleau-Ponty 2012, 131). Furthermore, in both the Husserlian and the hermeneutic traditions, our sedimented typifications take the form of prejudices, or pre-judgments (Taipale 2015; Fernandez 2016). While the everyday notion of prejudice has predominantly negative connotations, the phenomenological account is largely neutral (Gadamer 2008). Our background prejudices, or biases, are a necessary condition for successfully navigating our environment and appropriately typifying our objects of experience.

Fifth, the particular typifications we employ when making sense of an object of experience are responsive or adaptive to new experiences. While an object might immediately show up to me as a certain type, it can potentially reveal itself as something else. This is not to say that wholly novel typifications suddenly arise in light of new experiences. Rather, the typification we employ might be challenged, and another sedimented typification arises to take its place. Merleau-Ponty illustrates this process with an example of walking down a beach and seeing a grounded ship in the distance. From a certain angle, the mast merges with the forest, indistinguishable from the trees. But as he rounds the corner, the mast suddenly reunites with the ship.

> As I approached, I did not perceive the resemblances or the proximities that were, in the end, about to reunite with the superstructure of the ship in an unbroken picture. I merely felt that the appearance of the object was about to change, that something was imminent in this tension, as a storm is imminent in the clouds. The spectacle was suddenly reorganized, satisfying my vague expectation. (Merleau-Ponty 2012, 17–18)

The object now appears to him *as* a mast, and certain features of it are now obvious to him. He stresses, however, that this is not a cognitive or intellectual endeavor. It is illustrative of some more foundational aspect of experience.

Sixth, the typifications that structure our experience are fuzzy. They have fluid or ambiguous boundaries, which might be challenged or modified in light of new experiences. Suppose, for example, that I have only ever come across chairs with four legs. I am then presented with an object that seems to function as a chair (e.g., seats a single person, has a back, etc.), but stands on three legs. In such a case, I might experience it *as* a chair, thereby broadening the typification. Alternatively, the object might appear too different from what I know a chair to be, thereby establishing itself as some other kind of seat. There are not hard and fast rules for which of these events might occur, and the possibility of such experiences reveals that our typifications are not derived from clearly defined concepts.

Psychiatric Typification

In light of this initial sketch, it is clear that the typifications we employ in everyday experience are in no way scientific. They are, for the most part, tacit or implicit. They sediment and build up through life experiences without being questioned. And they do not have distinct boundaries or explicit criteria for what does or does not count as the type in question. In light of their largely unscientific nature, it is natural to ask why phenomenologists believe that typifications have something to contribute to psychiatric practice.

At least for Schwartz and Wiggins, the primary appeal of typifications is their ability to explain one of the "mysteries of psychiatric diagnosis" (Schwartz and Wiggins 1987b). The mystery they have in mind is the psychiatrist's ability to rapidly diagnose a patient, sometimes even by appeal to what is clearly subjective, first-personal criteria, such as the praecox feeling in the diagnosis of schizophrenia.[3] As they explain, many psychiatrists arrive at a diagnosis within the first few minutes of an interview (and sometimes in as little as 30 seconds). While additional interview time and information might alter the diagnosis, in the majority of cases it does not; the initial, rapid diagnosis is maintained (Gauron and Dickinson 1966a; 1966b; 1969; Kendell 1975; Sandifer Jr., Hordern, and Green 1970). In addition, they point to evidence that psychiatrists might be unaware of the criteria they appeal to when performing a diagnosis and that the criteria they do rely on sometimes differ from those they believe they are employing (Schwartz and Wiggins 1987b, 66). Many of these conclusions have been corroborated in more recent studies of rapid diagnosis and the use of intuition (Pallagrosi et al. 2016; Srivastava and Grube 2009; Witteman, Spaanjaars, and Aarts 2012). In at least one study, it was even found that psychiatrists who exhibited higher rationality,[4] as well as those who spent more time thinking about a prototypical case, were less accurate in their diagnoses than those who relied on their intuition (Aarts et al. 2012).

As Schwartz and Wiggins admit, the ability to categorize and diagnose a patient in such a short period of time, and without explicitly appealing to particular criteria, seems decidedly unscientific. However, this is precisely what they aim to demystify by their appeal to typification. They claim that once the process of typification is properly understood, the mystery will disappear, and "the truly *objective* and *scientific* status of psychiatric diagnosis can be explained and justified" (Schwartz and Wiggins 1987b, 69).

While they admit that *everyday* typifications are by no means scientific, they argue that this capacity can be supplemented with a "critical attitude." This is a stance in which we allow ourselves to doubt what we immediately experience something *as* (Schwartz and Wiggins 1987b, 72). By employing this critical attitude, we doubt the meaningful structure of our experienced object, leading us to seek out disconfirming evidence. As Schwartz and Wiggins put it, "Typifications are scientific *only to the extent that they are based upon and tested by evidence*" (1987b, 73).

As illustrated above, the everyday typification we tacitly apply in a certain experience might be replaced by another typification in light of future experiences. We do not typically seek such experiences out. But when we do have a disconfirming experience, our typification is nevertheless undermined and revised. A scientific approach to typification is one that harnesses this process of revision, making this tacit element of typification explicit and actively controlled by the experiencing subject.

IDEAL TYPES

While the process of typification, modified through a critical attitude, offers a justification for rapid and intuitive *diagnosis* in psychiatry, it does not yet provide a full-fledged approach to psychiatric *classification*. At most, it guides the practice of diagnosis without making any further claims about what kind of classificatory system these diagnostic categories belong to. Schwartz and Wiggins, however, want to make as much use of a scientifically supplemented typification as possible, and do so by establishing an ideal type approach to psychiatric classification. This approach, they argue, avoids a number of pitfalls inherent in the operational approach of the *DSM* and other traditional classificatory systems, such as the *International Classification of Diseases (ICD)*. Some of these approaches, as well as their relation to ideal types, are discussed in section three.

In order to clarify the notion of ideal types, I first briefly address their origin in the work of Max Weber and Karl Jaspers. Following this, I articulate the nature and role of ideal types in psychiatric practice.

The Origin of Ideal Types

While Schwartz and Wiggins draw the notion of typification from the phenomenological tradition, especially the work of Husserl and Schutz, they draw the notion of ideal types from Weber, as well as Jaspers' adaptation in *General Psychopathology* (1997; originally published in 1913). According to Weber, ideal types fill a void between what he calls the nomological and the idiographic modes of inquiry. The former investigates universals (e.g., natural laws), while the latter investigates particulars (e.g., individual human beings). However, in establishing the field of sociology, Weber realized that much of his subject matter does not clearly fit in either the nomological or the idiographic camp, instead occupying an uncomfortable middle ground. He was clearly interested in social practices and traditions, rather than particular historical figures or events. But these practices and traditions do not lend themselves to straightforward natural-scientific definition or the traditional conceptual clarifications of philosophy. They cannot be adequately articulated by appealing to necessary laws in the nomological mode of inquiry. Instead, there are core features that are more or less definitive of the phenomenon in question, but might not hold in every instance. In light of this, Weber claimed that the concepts that frame and guide his investigations are idealized, offering a clear definition while admitting that real instances of the phenomenon are unlikely to manifest all of the core features of the ideal type. As he says,

> An ideal type is formed by the one-sided *accentuation* of one or more points of view and by the synthesis of a great many diffuse, discrete, more or less present and occasionally absent *concrete individual* phenomena, which are arranged according to those one-sidedly emphasized viewpoints into a unified *analytical* construct. In its conceptual purity, this mental construct cannot be found empirically anywhere in reality. It is a *utopia*. (Weber 1949, 90)

Ideal types, according to Weber, are heuristic devices developed by appealing to our specific needs. We accentuate certain features insofar as they are the kinds of features our scientific investigation is concerned with. Therefore, ideal types were never meant to conceptualize the objective world. They allow us to conceptualize and delineate the value-laden world of culture, history, and human subjectivity.

Jaspers, writing a few years after Weber's 1904 introduction of ideal types in "'Objectivity' in Social Science and Social Policy," found the notion of ideal types useful for conceptualizing psychiatric disorders that were not yet clearly delineated, such as personality disorders. In the absence of well-defined categories, ideal types could be employed as initial heuristic devices, offering a shared conceptual framework for diagnosis and research. However, in light of the fact that most of the current *DSM* categories still have poorly defined boundaries (as evidenced by high rates of comorbidity and lack of distinct biological markers), more recent proponents of the ideal type approach employ these types across the full range of mental illness, including in studies of depression and schizophrenia.

Ideal Types in Contemporary Psychiatry

In the hands of Schwartz and Wiggins, ideal types offer an alternative to the dominant approach to psychiatric classification found in the *DSM-III* and later editions. While Weber and Jaspers claimed that ideal types can guide idiographic inquiry, Schwartz and Wiggins argue that they can also guide nomological inquiry. For example, psychiatrists might rally around a proposed ideal type in order to have a shared starting point for research—seeking out, for example, distinct neurobiological substrates or testing treatment response across a population with common psychopathological characteristics. In the case of idiographic inquiry, by contrast, psychiatrists can employ a shared set of diagnostic categories to initially characterize their patients, while admitting that this is only a starting point for more personalized therapeutic interventions.

In ideal types, Schwartz and Wiggins see the potential for a system of classification inspired by the psychiatrist's intuitive grasp of her patient. Rather than forcing psychiatrists to act as if their skilled understanding can be distilled into the checklist diagnostic methods required by the *DSM*, they take the diagnostic approach that comes most naturally to the clinician and construct it into something genuinely scientific.

In order to clarify the nature and role of ideal types in Schwartz and Wiggins's account, it will be helpful to contrast them with everyday typifications. The primary point of distinction is that ideal types make explicit what typifications leave tacit—and in a way that extends beyond their initial proposal for scientific typifications. Typifications are preconceptual, structuring the meaningful field of experience. We remain largely unaware of the criteria being employed when we typify objects in everyday experience. Ideal types, in contrast, focus in on a few representative criteria, or central aspects of the phenomenon. As Schwartz and Wiggins characterize them, "Ideal types are ... idealized definitions of typifications" (Schwartz and Wiggins 1987a, 282). They require that we draw out the features that are characteristic of the type in question and use these to articulate clear and explicit definitions.

In addition, ideal types, like typifications, are developed within particular contexts (e.g., the field of mental health) and for the sake of particular goals (e.g., accurately targeting therapeutic interventions or isolating neurobiological substrates). When constructing ideal types, these contexts and goals are made explicit. By clearly articulating our goals, we can actively select which features to highlight in our construction of the concept. By acknowledging that ideal types are constructed for the sake of achieving certain practical ends, it become clear that ideal types do not supply us with objectively valid categories. They are explicitly developed and employed as heuristic devices, only retained insofar as they help us achieve our aim.

Ideal types also build upon the anticipatory nature of typifications. However, rather than relying on the tacit anticipations or expectations that arise from everyday typifications, ideal types are developed with the aim of making accurate scientific predictions. Furthermore, these predictions might be made in either the nomological or the idiographic mode. A researcher might employ ideal types when subtyping depression, establishing preliminary distinctions that can be used to test treatment efficacy across the various subtypes. If those subtypes correlate with distinct treatment effects, then a clinician might use this set of ideal types for idiographic practice, such as diagnosing and treating individual patients.

In addition, while typifications are passed down and sedimented over a community's history, an ideal type approach loosens or disturbs this sediment. Rather than employing the taken for granted typifications we have acquired, including those that we have inherited from specialized subcommunities (e.g., the psychiatric community that assumes the legitimacy of *DSM* categories), the proponent of ideal types will critically assess the heuristic value of these types. Undermining the value of these typifications, she can propose new, and more clearly defined, ideal types to replace them. This allows the proponent of ideal types to disrupt and reorient her manner of typifying and categorizing her objects of inquiry, actively proposing new types that might help her better navigate her intellectual terrain.

Which ideal type we use to classify a particular person can also be modified in light of new experiences. However, in contrast with everyday typifications, the ideal type approach requires that we actively seek out disconfirming evidence to make certain that we have applied the best available ideal type. In this respect, the ideal type approach subsumes the scientific version of typification discussed above, integrating it into a broader classificatory system.

Ideal types are also discrete. Unlike everyday typifications, they do not have fuzzy or ambiguous boundaries. As Schwartz and Wiggins say,

> By idealized we mean descriptions that attempt to draw clear conceptual boundaries around features of things that, as directly given to us in immediate concrete situations, are ambiguous and unclear. ... In defining an ideal type we try to set aside ... indistinctness, ambiguity, and extreme variation and imagine a pure case in which the relevant features are distinct, unambiguous, and invariant. (1987a, 282)

This does not mean that ideal types can only be used in the categorization of phenomena that are themselves discrete. In Weber's initial introduction of ideal types, it is clear that his own objects of study lacked discrete boundaries. Ideal types are employed because of, not in spite of, the messy and indistinct nature of our subject matter. They do not have exact correlates in reality. It is assumed that the concrete phenomena we categorize with the ideal type will only meet some of the core criteria.

Finally, as mentioned above, ideal types are heuristic devices. They do not delineate true or objective kinds. Rather, they are employed in precisely those cases where we do not yet have an adequate grasp on where the boundaries between phenomena lie, or whether such boundaries exist at all. In the absence of such information, we nevertheless require some framework for shared discourse and scientific inquiry. Ideal types provide this framework while simultaneously initiating a self-reflective attitude in which we admit that our categorial distinctions are merely heuristic devices. This stance provides a framework for scientific inquiry while promoting the revision and refinement of the framework itself.

RESOLVING CONFUSIONS

Having clarified the nature of ideal types and their role in psychiatry, I here resolve some common confusions in the contemporary literature. These confusions often arise when ideal types are compared with other approaches to psychiatric classification, including prototypical, monothetic, and polythetic approaches.[5] Some of these confusions stem from apparently contradictory remarks made by Schwartz and Wiggins themselves. For example, in some works they claim that ideal types can subsume or conceptually unify some of these other approaches (Wiggins and Schwartz 1994, 96–97). In other works, by contrast, they argue that ideal types should displace these other approaches (e.g. Schwartz, Wiggins, and Norko 1995; Wiggins and Schwartz 1991). Further mischaracterizations are found in the work of other phenomenologists who attempt to employ ideal types in their own investigations. In order to adequately clarify the nature of ideal types as well as their relation with these alternative approaches, I here consider the prototypical, monothetic, and polythetic systems of classification in turn.

Prototypical Approaches

Schwartz, Wiggins, and Michael Norko draw their version of prototypes from the work of Eleanor Rosch (1973), who argued that many of our basic concepts are formed by engaging with representative exemplars of the category in question, rather than by establishing necessary and sufficient conditions. As they point out, ideal types and prototypes share a number of central traits. Both recognize the fuzziness of actual cases of disorder and admit their variable manifestations. Both aim to overcome this fuzziness by developing precise definitions of the disorders in question. And both aim to overcome variability by admitting that whether an individual case fits a type is always a matter of degree (Schwartz, Wiggins, and Norko 1989, 5; 1995, 426).

In spite of these similarities, they claim that prototypes and ideal types differ in at least three respects. As I argue, however, these differences are questionable, and may not accurately reflect the nature of prototypes as they are employed today.

First, echoing Jaspers, they claim that prototypes are composed of a list of features without any conceptual unity. Ideal types, in contrast, always aim to articulate some inherent order or unity in the apparently diverse features (Schwartz, Wiggins, and Norko 1989, 5–6; 1995, 426). However, in light of contemporary phenomenological work, this characterization of prototypes is misleading. Parnas and Gallagher, for example, have recently defended a prototype approach to psychiatric diagnosis (Parnas and Gallagher 2015). One of the reasons they do so is precisely because prototypes, unlike *DSM* categories, offer the same conceptual unity that Schwartz, Wiggins, and Norko attribute to ideal types. Their approach articulates the organizing principle at the heart of the disorder. They articulate what phenomenologists refer to as "motivational relations" among the various component features of a disorder, examining how one feature might bring about another, or why certain features tend to arise together. By appealing to a prototype approach, Parnas and Gallagher show that prototypical diagnosis can pick up on the gestalt organizations that are characteristic of psychiatric disorders. In short, while Schwartz and Wiggins's contrastive account might be reflective of the prototypes Jaspers was familiar with, it does not represent the current usage, especially among phenomenologists.

Second, Schwartz and Wiggins argue that a prototype approach is not in a position to produce new categories. This is because we construct prototypes by surveying psychiatrists on how they think of various disorders. The results of these surveys are amalgamated into the prototypical description of each category of disorder. Such a method, as Schwartz and Wiggins point out, can only tell us how we already conceive of the field of mental disorders. It cannot provide anything new (Schwartz, Wiggins, and Norko 1989, 6; 1995, 427–8). Ideal types, in contrast, are specifically designed to produce novel categories. While the process for producing these novel categories is not worked out in considerable detail, they point to the psychiatrist's ability to specify how particular cases diverge from the ideal. These divergences produce questions and hypotheses for further research, and might instigate revisions or require the production of new ideal types.

While the aim of producing a classificatory system that is dynamic and responsive to new evidence is commendable, it is not clear that prototypes cannot achieve the same end. A commitment to such a classificatory system is not necessarily alien to the prototype approach, and prototypes do not necessarily have to be drawn from surveys of psychiatrists' current understandings of disorders. At least on the account proposed by Parnas and Gallagher, psychiatrists should take seriously the project of developing and specifying prototypes that help us accurately diagnose and treat patients.

Third, Schwartz, Wiggins, and Norko argue that prototypes only provide us with the best example of a type, leaving out all of the atypical instances of the disorder. Ideal types, in contrast, offer us a concept that encompasses all instances of the category in question, thereby subsuming prototypes as just one of the possible manifestations of the condition included in the broader ideal type (Schwartz, Wiggins, and Norko 1995, 425–6). This distinction would certainly provide a reason to prefer ideal types to prototypes. However, the distinction seems to be based on misleading characterizations of both prototypes and ideal types. Prototypes do encompass the broad range of atypical manifestations of a disorder. They admit that the prototype is simply an exemplar and that actual instances of the disorder are likely to diverge from the prototype in various respects. Furthermore, ideal types, by Schwartz and Wiggins's own account, do not encompass the diversity of a disorder by offering a concept that is broad enough to encompass all of its manifestations. Rather, ideal types encompass the diverse range of manifestations in the same manner as prototypes. Proponents of ideal types admit that concrete instances of the disorder will not match the ideal type in all respects, so the psychiatrist must be willing to include instances of the disorder that are not fully encompassed by the neatly defined concept. In short, this wedge that Schwartz, Wiggins, and Norko attempt to drive between prototypes and ideal types seems forced or, at the very least, based on outdated notions of prototypes.

Finally, I want to suggest one respect in which prototypes have an advantage over ideal types. At least on Weber's account, we will never find a concrete instance of the ideal type. It is, as he calls it, a "utopia." Prototypes, by contrast, are defined through a concrete exemplar. In short, an ideal type is organized around an abstract concept, while a prototype is organized around a concrete condition. In light of this, prototypes have an edge when developing a firsthand approach to teaching and training—an approach that Schwartz and Wiggins are themselves supportive of (Schwartz and Wiggins 1987b). The trainee can be guided through the process of diagnosis by engaging with patients who exhibit the prototypical condition. In the case of ideal types, by contrast, each concrete condition will be in some respect imperfect, and this imperfection will have to be noted throughout the training process.

Monothetic Approaches

Monothetic approaches classify phenomena by appealing to a set of essential features—that is, those features that must hold for the phenomenon to count as the kind of phenomenon that it is. As Schwartz and Wiggins explain, monothetic approaches are antithetical to the ideal type approach. The criteria that define an ideal type consist in only those features that are *typical* or *representative*—not *essential*. As Schwartz and Wiggins argue, concrete cases of a disorder are unlikely to meet all of the criteria of the ideal type, but they are nonetheless instances of this type. In fact, if we take seriously Weber's claim that ideal types are utopias, having no exact correlate in the real world, they could not possibly be defined in terms of necessary criteria. There would be no concrete phenomena capable of meeting all the essential criteria, and thus no concrete phenomena that could be legitimately categorized by appealing to the ideal type.

Despite Schwartz and Wiggins's distinction between ideal types and monothetic classifications, the two approaches have still been confused in the phenomenological literature. Parnas and Zahavi, for example, have argued that phenomenological psychopathologists aim to uncover the essential features of a disorder. In at least one case, they align this approach with the ideal types developed by Weber and popularized by Schwartz and Wiggins.

They conflate ideal types with essences, or sets of essential features, claiming that the "ideal type exemplifies the *ideal* and *necessary* connections between its composing features" (Parnas and Zahavi 2002, 157).

However, in spite of their mischaracterization of ideal types, it is important to note that Parnas and Zahavi are not mischaracterizing phenomenology's role in psychiatric classification. The monothetic system that they sketch has its roots in the Husserlian tradition, which articulates the essential structures of human subjectivity. When applied to the domain of psychopathology, this approach requires that we articulate the essential features of particular psychopathological conditions, drawing discrete boundaries and supplying necessary criteria for diagnosis.

This approach still finds support among phenomenologists today, and for good reason. If phenomenologists were able to pin down essential features of psychopathological subjectivity—whether these be features of a particular symptom, or features of a category as a whole—this would likely be a boon to psychiatric research and diagnosis. Such criteria would provide unambiguous ways of distinguishing among various conditions, increasing, for example, the clinician's ability to accurately target therapeutic interventions and the researcher's ability to isolate distinct biomarkers. While it is important that we do not confuse such an approach with ideal types, it should be kept in mind that the appeal to essential features can be accommodated by certain phenomenological traditions, and might offer a competing phenomenological approach to psychiatric classification.

Polythetic Approaches

In contrast with monothetic approaches, polythetic approaches classify phenomena by appealing to a set of criteria, only some of which have to be met for the phenomenon to be included within the category. As Schwartz and Wiggins point out, the operational approach of the *DSM-III* and subsequent editions is the most well-known example of a polythetic classification in psychiatry. However, in spite of Schwartz and Wiggins's insistence that ideal types offer an alternative to the polythetic approach of the *DSM*, there are still misunderstandings over the relationship between ideal types and the *DSM* in the contemporary literature.

Ghaemi, for example, has claimed that the *DSM* categories can be thought of as ideal types (Ghaemi 2003; 2006). It is true that *DSM* categories share at least one important feature with ideal types. *DSM* categories were developed as heuristic devices, meant to guide clinical practice and be further refined through continued research. As is now well known, the initial *DSM-III* categories reified, many remaining largely unchanged in the *DSM-IV* and *DSM-5* in spite of evidence that many of the categories include a heterogeneity of conditions. Nevertheless, insofar as the *DSM* remains theoretically neutral on the causes of disorders, it continues to offer a broadly heuristic outlook, even if its heuristic devices are not especially useful.

Ideal types and *DSM* categories, however, share little beyond this heuristic element. While ideal types focus only on the central features of a disorder, *DSM* categories are delimited by reference to operational, or easily observable, criteria. Such criteria *might* be central or defining characteristics of the disorder (e.g., depressed mood in a diagnosis of major depressive disorder), but in many cases they are not. Furthermore, diagnoses conducted with ideal types specify how closely the particular instance of the disorder matches its ideal counterpart, thereby integrating a dimensional component into diagnosis. In the *DSM* system, by contrast, diagnosis is binary. A particular condition either does or does not meet the required number of criteria (although in some cases a psychiatrist might note that a patient exhibits a sub-threshold version of a disorder, having just under the required number of symptoms).

Having clarified some of the misunderstandings surrounding the notion of ideal types, there remains the question of why Schwartz and Wiggins sometimes claim that ideal types can subsume or unify other approaches (especially prototypical and polythetic approaches), while in other cases they argue that ideal types should displace these other approaches. I think that the resolution to this apparent contradiction is fairly straightforward. While *traditional* prototypical and polythetic approaches to classification do not share much in common with ideal types, they are amenable to modifications that might bring them closer to ideal types. Polythetic approaches might focus only on central features of the disorder, and they might be thought of as heuristic devices, rather than as objectively valid categories. Furthermore, prototypes, as pointed out above, might posit a conceptual unity among their component features, and can also be thought of as heuristic devices.

However, while such approaches will certainly benefit from borrowing elements from ideal types, and will therefore share much in common with them, it does not follow that they will thereby *become* ideal types. For example, a prototype's orientation around a concrete condition rather than an abstract concept means that it cannot fully merge with ideal types. This distinctive element of prototypes conflicts with a basic feature of ideal types. It is likely that each classificatory approach will retain some distinctive elements that cannot be fully reconciled with ideal types. In short, while other classificatory approaches can be brought closer to ideal types, it does not follow that ideal types can conceptually unify these other approaches, or fully subsume them.

CONCLUSION

In this chapter, I have offered a critical introduction to typification and ideal types in phenomenological psychiatry. As explained above, the dominant approach to psychiatric classification employed in the *DSM-III* and later editions fails to exhibit certain features that should be desirable in a classificatory system. First, the polythetic, operational approach of the *DSM* defines disorders by appealing to sets of easily observable symptoms. In most cases, no attempt is made to understand why these symptoms tend to arise together, or why certain typical symptoms fail to manifest in particular cases. Ideal types, in contrast with the polythetic approach of the *DSM*, focus only on those criteria deemed central to the disorder (rather than prioritizing those that are easily observable). In so doing, ideal types (and, on some accounts, prototypes) put us in a better position to understand how these central features are organized and related, granting us greater insight into the category itself.

Second, by requiring that diagnoses be conducted through structured interviews and what is essentially a checklist of atomistic criteria, the *DSM* forces the clinician to rely on a nonintuitive understanding of her patient. As Schwartz and Wiggins argue, we tacitly understand and categorize our objects of experience, including other human beings, by employing typifications. Even when a clinician uses the checklist diagnostic approach of the *DSM*, she cannot avoid making sense of her patient through tacit typifications. In light of this, Schwartz and Wiggins, along with many other phenomenologists, believe that we would be better off with a classificatory system that harnesses this everyday capacity. This does not mean that clinicians should be encouraged to diagnose based on gut instinct. Rather, it requires that the clinician actively test her initial intuition (rather than ignore it), seeking out disconfirming evidence until she arrives at an accurate diagnosis.

Third, and finally, the *DSM* approach has produced a set of reified categories. In spite of continued advances in psychiatric research and substantial evidence of the heterogeneity within certain categories of disorder, the set of categories devised in the *DSM-III* (as well as their diagnostic criteria) remain largely unchanged in the *DSM-5*. One reason for this reification is that we have assumed the legitimacy of the *DSM* categories. They frame both our scientific and everyday discourse, making it difficult to think outside of them. According to Schwartz and Wiggins, ideal types remedy the problem of reification by explicitly establishing themselves as heuristic devices. The proponent of ideal types makes no claims to the factual accuracy or objective validity of his distinctions. He provides them as tools for orienting the psychiatric community toward common ends. Insofar as they help us achieve these ends, they are retained. But if we find that they provide no heuristic value, we can simply revise or dispense with them. We hold no commitment to our ideal types that extend beyond their utility.

In conclusion, ideal types offer an alternative approach to psychiatric classification and diagnosis—one that promises to remedy many of the problems that psychiatry faces today. However, the greatest strength of the ideal type approach is found not in the particular remedies it offers, but in a new, self-reflective attitude that permeates the approach as a whole. This self-reflective attitude catalyzes the revision and modification of the types we employ in research and practice, producing a psychiatric program that is responsive to its own evidence and aware of its own needs. It supplies a foundation that is revisable from the ground up. This approach admits its own imperfections for the sake of producing something better.

NOTES

1. This foundational relationship between typification and ideal types is somewhat anachronistic. Ideal types were proposed by Weber as early as 1904 in "'Objectivity' in Social Science and Social Policy," and were later adapted by Jaspers in *General Psychopathology*, first published in 1913. The notion of typification, at least as employed by Schwartz and Wiggins, stems from Husserl's work on passive synthesis developed in the 1920s (1973; 2001). In short, ideal types were already developed and employed well before we had an adequate phenomenological account of typification. However, this historical order of development does not contradict Schwartz and Wiggins's claim that ideal types rely upon our everyday manner of experiencing, and that to properly understand ideal types, we first need to understand some of the constitutive features of human experience.

2. Following Schwartz and Wiggins, I reserve the term "type" for discussions of ideal types and prototypes. "Typifications," in contrast, are the structural features of everyday experience upon which "types" are founded.

3. In their discussion, Schwartz and Wiggins technically point to *two* mysteries of psychiatric diagnosis: rapid diagnosis and the praecox feeling. However, I take the praecox feeling to be a well-established example of rapid diagnosis based on subjective feelings—in this case, feelings of poor rapport between the psychiatrist and the patient that are indicative of the dementia praecox, or schizophrenia. For this reason, I do not treat it separately.

4. "Rationality," as employed in the study, is understood as a style of thinking that can be measured by the rational-experiential inventory (REI). As Pacini and Epstein describe it, "The rational system is an inferential system that operates by a person's understanding of culturally transmitted rules of reasoning; it is conscious, relatively slow, analytical, primarily verbal, and relatively affect-free" (Pacini and Epstein 1999, 972).

5. Schwartz and Wiggins also discuss the relationship between ideal types and dimensions in some of their work (Wiggins and Schwartz 1991). However, I here

confine my discussion to categorical classifications. For a discussion of phenomenology and recent developments in dimensional systems of research and classification, see Fernandez (forthcoming).

REFERENCES

Aarts, Alexander A., Cilia L. M. Witteman, Pierre M. Souren, and Jos I. M. Egger. 2012. "Associations between Psychologists' Thinking Styles and Accuracy on a Diagnostic Classification Task." *Synthese* 189 (S1): 119–30.

Cuthbert, Bruce N. 2014. "The RDoC Framework: Facilitating Transition from ICD/DSM to Dimensional Approaches That Integrate Neuroscience and Psychopathology." *World Psychiatry* 13 (1): 28–35.

Cuthbert, Bruce N., and Thomas R. Insel. 2013. "Toward the Future of Psychiatric Diagnosis: The Seven Pillars of RDoC." *BMC Medicine* 11 (1): 126.

Fernandez, Anthony Vincent. forthcoming. "Phenomenology and Dimensional Approaches to Psychiatric Research and Classification." *Philosophy, Psychiatry, & Psychology*.

Fernandez, Anthony Vincent. 2016. "The Subject Matter of Phenomenological Research: Existentials, Modes, and Prejudices." *Synthese*. doi: 10.1007/s11229-016-1106-0.

Gadamer, Hans-Georg. 2008. *Philosophical Hermeneutics*, edited by David E. Linge. Berkeley, CA: University of California Press.

Gauron, Eugene F., and John K. Dickinson. 1966a. "Diagnostic Decision Making in Psychiatry: II. Diagnostic Styles." *Archives of General Psychiatry* 14 (3): 233.

Gauron, Eugene F., and John K. Dickinson. 1966b. "Diagnostic Decision Making in Psychiatry: I. Information Usage." *Archives of General Psychiatry* 14 (3): 225–32.

Gauron, Eugene F., and John K. Dickinson. 1969. "The Influence of Seeing the Patient First on Diagnostic Decision Making in Psychiatry." *American Journal of Psychiatry* 126 (2): 199–205.

Ghaemi, S. Nassir. 2003. *The Concepts of Psychiatry: A Pluralistic Approach to the Mind and Mental Illness*. Baltimore, MD: Johns Hopkins University Press.

Ghaemi, S. Nassir. 2006. "Feeling and Time: The Phenomenology of Mood Disorders, Depressive Realism, and Existential Psychotherapy." *Schizophrenia Bulletin* 33 (1): 122–30.

Husserl, Edmund. 1973. *Experience and Judgment*. Evanston, IL: Northwestern University Press.

Husserl, Edmund. 2001. *Analyses Concerning Passive and Active Synthesis: Lectures on Transcendental Logic*. Boston, MA: Springer.

Insel, Thomas. 2013. "Director's Blog: Transforming Diagnosis." April 29. http://www.nimh.nih.gov/about/director/2013/transforming-diagnosis.shtml.

Jaspers, Karl. 1997. *General Psychopathology*. Translated by J. Hoenig and Marian W. Hamilton. Baltimore, MD: Johns Hopkins University Press.

Kendell, Robert E. 1975. *The Role of Diagnosis in Psychiatry*. Philadelphia, PA: Wiley-Blackwell.

Merleau-Ponty, Maurice. 2012. *Phenomenology of Perception*. Translated by Donald Landes. New York, NY: Routledge.

Pacini, Rosemary, and Seymour Epstein. 1999. "The Relation of Rational and Experiential Information Processing Styles to Personality, Basic Beliefs, and the Ratio-Bias Phenomenon." *Journal of Personality and Social Psychology* 76 (6): 972.

Pallagrosi, Mauro, Laura Fonzi, Angelo Picardi, and Massimo Biondi. 2016. "Association between Clinician's Subjective Experience during Patient Evaluation and Psychiatric Diagnosis." *Psychopathology* 49 (2): 83–94.

Parnas, Josef, and Dan Zahavi. 2002. "The Role of Phenomenology in Psychiatric Diagnosis and Classification." In *Psychiatric Diagnosis and Classification*, edited by Mario Maj, Wolfgang Gaebel, Juan José López-Ibor, and Norman Sartorius, 137–62. New York, NY: John Wiley & Sons.

Parnas, Josef, and Pierre Bovet. 2015. "Psychiatry Made Easy: Operation(al)ism and Some of Its Consequences." In *Philosophical Issues in Psychiatry III: The Nature and Sources of Historical Change*, edited by Kenneth Kendler and Josef Parnas, 190–212. Oxford: Oxford University Press.

Parnas, Josef, and Shaun Gallagher. 2015. "Phenomenology and the Interpretation of Psychopathological Experience." In *Re-Visioning Psychiatry: Cultural Phenomenology, Critical Neuroscience, and Global Mental Health*, edited by Laurence J. Kirmayer, Robert Lemelson, and Constance A. Cummings, 65–80. New York, NY: Cambridge University Press.

Polanyi, Michael. 1966. *The Tacit Dimension*. Chicago, IL: University of Chicago Press.

Polanyi, Michael. 1974. *Personal Knowledge: Towards a Post-Critical Philosophy*. Chicago, IL: University of Chicago Press.

Ratcliffe, Matthew. 2015. *Experiences of Depression: A Study in Phenomenology*. Oxford: Oxford University Press.

Rosch, Eleanor H. 1973. "On the Internal Structure of Perceptual and Semantic Categories." In *Cognitive Development and the Acquisition of Language*, edited by Timothy E. Moore, 111–43. Oxford: Academic Press.

Sandifer Jr, Myron G., Anthony Hordern, and Linda M. Green. 1970. "The Psychiatric Interview: The Impact of the First Three Minutes." *American Journal of Psychiatry* 126 (7): 968–73.

Sass, Louis, and Elizabeth Pienkos. 2013. "Varieties of Self-Experience: A Comparative Phenomenology of Melancholia, Mania, and Schizophrenia, Part I." *Journal of Consciousness Studies* 20 (7–8): 103–30.

Schutz, Alfred. 1966. "Type and Eidos in Husserl's Late Philosophy." In *Collected Papers III: Studies in Phenomenological Philosophy*, edited by Aron Gurwitsch, 92–115. The Hague: Martinus Nijhoff.

Schwartz, Michael A., and Osborne P. Wiggins. 1986. "Logical Empiricism and Psychiatric Classification." *Comprehensive Psychiatry* 27 (2): 101–14.

Schwartz, Michael A., and Osborne P. Wiggins. 1987a. "Diagnosis and Ideal Types: A Contribution to Psychiatric Classification." *Comprehensive Psychiatry* 28 (4): 277–91.

Schwartz, Michael A., and Osborne P. Wiggins. 1987b. "Typifications: The First Step for Clinical Diagnosis in Psychiatry." *The Journal of Nervous and Mental Disease* 175 (2): 65–77.

Schwartz, Michael A., Osborne P. Wiggins, and Michael A. Norko. 1989. "Prototypes, Ideal Types, and Personality Disorders: The Return to Classical Psychiatry." *Journal of Personality Disorders* 3 (1): 1–9.

Schwartz, Michael A., Osborne P. Wiggins, and Michael A. Norko. 1995. "Prototypes, Ideal Types, and Personality Disorders: The Return to Classical Phenomenology." In *The DSM-IV Personality Disorders*, edited by W. John Livesley, 417–32. New York, NY: Guilford Press.

Srivastava, Anil, and Michael Grube. 2009. "Does Intuition Have a Role in Psychiatric Diagnosis?" *Psychiatric Quarterly* 80 (2): 99–106.

Taipale, Joona. 2015. "From Types to Tokens: Empathy and Typification." In *The Phenomenology of Sociality: Discovering the "We,"* edited by Thomas Szanto and Dermot Moran, 143–58. London: Routledge.

Weber, Max. 1949. *Methodology of Social Sciences*, edited by Henry A. Finch and Edward A. Shils. Glencoe, IL: The Free Press.

Wiggins, Osborne P., and Michael A. Schwartz. 1991. "Research into Personality Disorders: The Alternatives of Dimensions and Ideal Types." *Journal of Personality Disorders* 5 (1): 69–81.

Wiggins, Osborne P., and Michael A. Schwartz. 1994. "The Limits of Psychiatric Knowledge and the Problem of Classification." In *Philosophical Perspectives on Psychiatric Diagnostic Classification*, edited by John Z. Sadler, Osborne P. Wiggins, and Michael A. Schwartz, 89–103. Baltimore, MD: Johns Hopkins University Press.

Witteman, Cilia L. M., Nanon L. Spaanjaars, and Alexander A. Aarts. 2012. "Clinical Intuition in Mental Health Care: A Discussion and Focus Groups." *Counselling Psychology Quarterly* 25 (1): 19–29. doi:10.1080/09515070.2012.655419.

Chapter 4

The Reproducibility[1] of Epistemic Humility

Abraham P. Schwab

Over the last 10 years, reproducibility has been demonstrated to be a serious concern for psychological science and medical science. Attempts to reproduce their published conclusions have produced mixed and sometimes dismal results. This chapter begins to answer an important practical and epistemological question raised by these results. How does the reproducibility problem for medical science, and to a lesser degree psychological science, affect the confidence medical practitioners[2] should have in their recommendations for patient care?

This chapter uses a 1991 study of oncology practitioners' overconfidence in their recommendations (Baumann, Deber, and Thompson 1991) to illustrate how the reproducibility problem affects how we should approach the risk of overconfidence in medical practitioners' recommendations. In short, the reproducibility problem exacerbates difficulties in addressing overconfidence among medical practitioners. A previously published taxonomy of epistemically humble medical judgments (Schwab 2012) attempted to address practitioner overconfidence by focusing on the distinction between the intuitive judgments of practitioners and the direct application of medical science's conclusions. But this Epistemic Humility Taxonomy paid inadequate attention to the quality of medical science's conclusions. The reproducibility problem illustrates that an adequate taxonomy should reference the robustness of these conclusions as a means to sift through medical science's conclusions.

One of the challenges that this chapter will not take up is the publication norms for psychological science and medical science. As Pashler and Harris (2012) put it, "There are likely to be serious replicability problems in our field, and correcting these errors will require many significant reforms in current practices and incentives." They, along with Ioaniddis (2005),

Trout (1993), and many others (see also Easterbrook et al. 1991), raise concerns about how incentives to publish, combined with the norm of not publishing reproductions or null results, affect the characteristics of published conclusions in medical and psychological science. While there are many important things to be said about the structure of incentives for publications and how these incentives contribute to the reproducibility problem, these questions will be left aside in favor of questions about the confidence medical practitioners should have in their recommendations. Even as the norms of publishing (and not publishing) set the stage for these recommendations, this chapter aims to identify strategies for practitioners to account for the reproducibility problem as they calibrate their confidence in their recommendations to patients.[3]

THE CASE OF OVERCONFIDENT
ONCOLOGY PRACTITIONERS

An early and valuable study at the intersections of medical and psychological science evaluated oncology practitioners' confidence in their treatment recommendations (Baumann, Deber, and Thompson 1991; moving forward referred to as the "BDT study"). The researchers presented cases to oncology physicians and nurses and asked these practitioners to make recommendations and estimate their confidence in the accuracy of their recommendations. Practitioners given similar cases recommended incompatible treatments with high degrees of confidence. The researchers dubbed this "micro-certainty vs. macro-uncertainty." Individual practitioners were sure they were recommending the best treatment (micro-certainty), even though, as a set, the practitioners recommended incompatible treatments for similar cases (macro-uncertainty).

The micro-certainty and macro-uncertainty findings of the DBT study provide some evidence of medical practitioner overconfidence. What's not clear from the study itself is the character of this overconfidence. The appropriate characterization of this overconfidence hinges on whether the conclusions of medical science provide clear conclusions about the best treatment recommendations. The implications of the "micro-certainty and macro-uncertainty" phenomenon depend on whether medical science has produced a clear conclusion about the best treatment to recommend.

Incompatible treatment recommendations from medical practitioners, like in the BDT study, are not a priori problematic. If the conclusions of medical science are unclear, disagreement among recommendations should be expected. There is no clearly defined best recommendation and macro-uncertainty is warranted. But, if the conclusions are unclear, the diverse recommendations should be coupled with low levels of confidence. Practitioners should recognize the lack of a clear conclusion and they should have low degrees of confidence in their recommendations. In the scenario where oncological science has not established a best treatment, all of the micro-certain practitioners in the BDT study are overconfident.

Alternatively, medical science may produce a clear conclusion regarding the best recommended treatment. In this scenario, practitioners who offer the best recommendation should have high degrees of confidence in their recommendations. High degrees of confidence for these practitioners is well calibrated, and micro-certainty is warranted. But macro-uncertainty contradicts the conclusions of medical science. Practitioners who recommend some other treatment are overconfident in their recommendation. They are overconfident simply because they make the wrong recommendation.[4]

Whether or not there is an established best treatment, the micro-certainty and macro-uncertainty identified in the BDT study indicate that at least some of the practitioners' confidence is poorly calibrated. When there is no clear conclusion from medical science, all practitioners with high degrees of confidence have poorly calibrated confidence in their recommendations. When there is a clear conclusion, but macro-uncertainty persists, some of the practitioners have poorly calibrated confidence because they are too confident in a recommendation that is contradicted by medical science. In either case, the BDT study indicates that at least some practitioners have poorly calibrated confidence in their recommendations.

Two basic strategies are clear contenders for improving poorly calibrated confidence and/or correcting mistaken judgments. First, education about medical science's conclusions could serve to calibrate practitioner confidence. When medical science demonstrates a clear conclusion about the best treatment, education would redirect existing (over)confidence to align with the clearly defined best treatment. Education, however, may be less successful when medical science has not demonstrated a best treatment. In those cases, education would focus on the lack of clear directives and, subsequently, on the limits of appropriate confidence in any recommendation. A well-calibrated confidence is low level of confidence. When medical science provides no clear conclusion, this strategy's success hinges on convincing practitioners that they should continue to make recommendations for patient care, but have limited confidence in those recommendations.

Second, practitioners may be prompted to recalibrate their confidence through a "consider the alternative" strategy. This is a simple and effective strategy for diminishing confidence (Koriat, Lichtenstein, and Fischoff 1980; Koehler 1991; Hirt and Markman 1995; Hirt, Kardes, and Markman 2004). When decision-makers "consider the alternative" to their judgment, when they consider that the best recommendation may be different from and incompatible with their recommendation, their confidence level lessens. In cases where the consensus of medical science is entirely unclear, this strategy would generally lead practitioners to better calibrated degrees of confidence. Applied too broadly, however, this strategy will have undesirable effects. Specifically, when practitioners have well-calibrated confidence in a clearly established (by medical science) best recommendation, diminishing these practitioners' confidence by asking them to "consider the alternative" could turn a well-calibrated high level of confidence into a poorly calibrated level of confidence. They should have high confidence, and asking them to consider the alternative may undermine that confidence.

The key to identifying the appropriate strategy for calibrating practitioner confidence hinges on identifying whether medical science clearly identifies a best treatment. The appropriate next step is to review the conclusions of medical science to establish whether there is a best treatment recommendation. One might expect that simply relying on the published conclusions of medical science would resolve this. It is here that the reproducibility problem is brought into sharp relief.

THE REPRODUCIBILITY PROBLEM

Discussing the methodologies common to medical and psychological science, Ioannidis (2005) provocatively argues that "most published research findings are false." He points to a number of contributing factors, including the norm of using a single study to declare a shift in a field's conclusions, the size of the field, the size of the study, and the possibility for bias from design and from financial relationships (2005). Ioannidis advocates for replication (or reproduction) of research conclusion as a mechanism to mark the distinction between "unconfirmed genuine discoveries and unchallenged fallacies" (Ioannidis 2012, 649). Without this, Ioannidis notes that "unchallenged fallacies may compromise the majority of circulating evidence" (645).

Ioannidis's concerns have gathered confirming support in both psychological science and medical science. In psychological science, the Open Science Collaboration (2015) began recreating 100 studies from top-tier journals to evaluate their reproducibility. Of the studies they recreated, only about 4 in 10 were judged as a successful reproduction, leaving open the possibility that more than half of the studies' conclusions are unchallenged fallacies.[5]

In medical science, attempts to evaluate reproducibility have returned even more dismal results. In preclinical research, Begley and Ellis (2012) attempted to replicate "landmark" preclinical oncology and hematology studies and the "scientific findings were confirmed in only 6 (11%) cases. Even knowing the limitations of preclinical research, this was a shocking result" (Begley and Ellis 2012, 532). In a study aimed to investigate the value of academic publishing for identifying marketable pharmaceuticals, Prinz, Schlange, and Asadullah (2011) could reproduce only 25% of published results.[6] Simon and Roychowdhury (2013) note how the press of personalized medicine in oncological practice will increase reliance on preclinical research: "It is difficult to apply this [Phase II and Phase III] model to patients with rare diseases or rare molecular subsets of certain types of cancer, for whom enrollment would be very low" (366). As narrowly defined applications of medical science increase, practitioners and patients will rely more heavily on preclinical medical science.

Concerns about reproducibility extend to the clinic as well. Doshi, Goodman, and Ioannidis (2013) discuss the possibilities of biased (and so not reproducible) publications about clinical trials. When practitioners and their patients put their trust in published literature for help making clinical decisions, they assume that the conclusions are not distorted and are based on complete information, but "the cases of Vioxx, Avandia, Celebrex, Paxil, Tamiflu, and Neurontin suggest that these assumptions are often unfounded" (645).

As noted in the previous section, if there is no clearly demonstrated best recommendation from medical science, the "consider the alternative" strategy is likely the best available means to calibrate practitioner confidence. If, however, there is a demonstrably best recommendation, education will generally be a preferable strategy for calibrating practitioner confidence and/or correcting mistaken applications of medical science. The reproducibility problem limits the ability to accurately distinguish between these kinds of cases, which in turn undermines the ability to better calibrate the (over)confidence of practitioners.

In an ideal circumstance, practitioners will have well-calibrated confidence in their recommendations for a particular condition. The problem of reproducibility suggests that (1) the published conclusions of medical science may not indicate the best recommendation and (2) practitioners will have difficulty identifying the appropriate levels of confidence in the recommendations based on the published conclusions of medical science.[7]

THE EPISTEMIC HUMILITY TAXONOMY

Previously, I sketched the Epistemic Humility Taxonomy (Schwab 2012) to frame how practitioners should calibrate their confidence in their applications of medical science. The Epistemic Humility Taxonomy includes the Defined Domain of application, the Ragged Edge of the Defined Domain, and the space beyond the Ragged Edge. As initially articulated, the various levels of the taxonomy correlate to the congruence between the features of medical science's conclusions and the features of the clinical situation at hand. For example, the congruence between a patient's characteristics and the inclusion and exclusion criteria of the relevant medical research indicate the appropriate categorization of a practitioner's judgment. When patient characteristics match the research subjects' characteristics, resulting applications of medical science fall within the Defined Domain. Practitioners should have high levels of confidence in these applications of medical science because the application falls neatly within the Defined Domain of the research itself. When patients have characteristics that match research subject characteristics to some degree, but also have complicating or confounding characteristics, applications of medical science fall along the Ragged Edge. This application of medical science should be less confident than within the Defined Domain, but more confident than recommendations without any support from relevant areas of medical science. When patient characteristics have no corollary research participant population, the applications of medical science are beyond the Ragged Edge. Here the judgments of practitioners will be better than the judgments of a lay person, who lacks the relevant background, but far less reliable than recommendations with direct support from medical science. Confidence in recommendations beyond the Ragged Edge should be low.

The Epistemic Humility Taxonomy rests on reliabilism[8] as the appropriate epistemic strategy for medical science and practice (Schwab 2008a; 2012). In the sense discussed here, reliabilism aims to identify processes that reliably predict future outcomes. The justification for a belief (and subsequent recommendation) comes from the reliability of the belief-forming process that led to said belief. This favors scientific inquiry over psychic interpretation. In medical practice, reliabilism privileges diagnostic tests that produce the fewest false negatives and false positives. In treatment, reliabilism privileges those treatments that demonstrate efficacy beyond the clinical trial over those with unproven effectiveness. In medical practice, reliabilism aims to identify the processes that reliably predict future outcomes. As I noted, "Claims produced by reliable processes are *exactly* what we look for in medical practice" (2008a, 304).

I have offered defenses of reliabilism elsewhere (Schwab 2008a; 2012) and will not review these defenses here. There are a couple of advantages of reliabilism worth mentioning. First, reliabilism favors accuracy over other epistemological characteristics (i.e., coherence) because reliabilist criteria sort processes according to their conduciveness to making accurate predictions. Even if a reliably produced belief runs counter to the usual practice or to commonly held beliefs, the reliability of the process is enough to justify it. In this way, reliabilism may unapologetically undermine existing beliefs and patterns of practice. If the process reliably makes accurate predictions, it does not need to be explained or integrated with common practice or commonly held beliefs.[9] While, all things considered, a rich understanding of all relevant aspects of a belief-producing process (including the best background theory of a judgment) may be preferred, such an understanding is not required for reliabilism. If a process reliably produces accurate predictions, that is enough. Second, reliabilism includes the flexibility to favor efficiency. If there are competing processes that are equivalently reliable, reliabilism allows us to pick the more efficient process. In this way, reliabilism favors precisely what medical science aims to achieve—the best recommendation given the constraints of time and resources.[10] In these ways, reliabilism is a particularly practical epistemological perspective for evaluating medical practitioners' judgment, recommendations, and confidence.

At this point, it helps to look at how medical science improves the quality of medical care. It does so in a couple of ways. First, it can improve practitioner understanding of the background conditions of medical practice by improving the accuracy of their views about the medical features of human beings. Medical science demonstrates the reliability of the general theories upon which medical practice is based. Second, and more germane to this chapter, medical science can provide clear and reliable directives about appropriate diagnostic tests and treatments for particular kinds of patients. Medical science provides guidance for narrowly defined decisions in medical practice.

The clear directives of medical science justify higher degrees of confidence in practitioner recommendations because they allow practitioners to avoid reliance on their intuitive judgment. I say this not to bury practitioners' intuitive judgment but to praise it. When medical science does not provide clear directives, practitioners are left to make complicated inferential judgments based on patient characteristics and the theoretical background of medical science. In the Epistemic Humility Taxonomy, intuitive judgments are made at the Ragged Edge or beyond. That practitioners make these judgments as well as they do is a truly impressive feature of human judgment. And in many cases, such intuitive judgments are the best available grounds for a recommendation. A decision must be made about what to recommend even when practitioners lack a clear directive from medical science. But, as I've discussed extensively elsewhere, recommendations by medical practitioners that require intuitive judgment are subject to the biases of human judgment (Schwab 2008b). Given this likelihood of bias when medical science does not provide clear directives, these judgments are less likely to be reliable. Accordingly, practitioners should have lower degrees of confidence in their recommendations based on their intuitive judgments than recommendations based on the conclusions of medical science. The Epistemic Humility Taxonomy distinguishes between the less reliable intuitive judgments of practitioners and the more reliable directives of medical science, and requires practitioners to acknowledge the differentials in confidence that attach to these distinct grounds for their recommendations.

Importantly, in its initial articulation, the Epistemic Humility Taxonomy was organized around the application of the published conclusions of medical science. But in this initial articulation, the Epistemic Humility Taxonomy failed to articulate any criteria for distinguishing *between* medical science's conclusions. It treated them as a uniform set. As the reproducibility problem demonstrates, this limits the accuracy and usefulness of the articulated applications of the taxonomy. Specifically, in order to help practitioners calibrate their confidence, the taxonomy should provide guidance for evaluating the published conclusions of medical science as well as the applications of those conclusions. To provide this guidance, the published conclusions of medical science should be evaluated according to criteria of robustness.

ROBUSTNESS, RELIABILISM, AND REPRODUCIBILITY

Robustness is not a new concept, nor is it one that should be entirely unfamiliar. Nonetheless, the literature analyzing the concept and its application is less, well, robust than ideal. So it's worth a quick sketch. Wimsatt (1981) characterizes robustness' general value as follows:

> All the variants and uses of robustness have a common theme in the distinguishing of the real from the illusory; the reliable from the unreliable; the objective from the subjective; the object of focus from the artifacts of perspective; and, in general, that which is regarded as ontologically and epistemologically trustworthy and valuable from that which is unreliable, ungeneralizable, worthless, and fleeting. (63)

In developing the features of robustness,[11] Wimsatt discusses two alternative ways to develop conclusions: as a chain of conclusions or as a set of conclusions (which might be called a "constellation" of conclusions, though Wimsatt declines to call it this). As a chain, conclusions draw upon a single perspective or set of foundational assumptions that ground the entire series of conclusions. Wimsatt notes the limits of a chain of conclusions. If the basis of any conclusion rests solely on the chain that precedes it, the conclusion will be as weak as the weakest conclusion in the chain. If a single conclusion in a chain is weak, every subsequent conclusion in that chain is at least that weak. If the grounding perspective turns out to be mistaken or a fundamental assumption is undermined, the whole chain of conclusions is undone.

As a set (or "constellation") of conclusions, independent assumptions or perspectives support related conclusions. As Wimsatt puts it, "With independent alternative ways of deriving a result, the result is always surer than its weakest derivation" (Wimsatt 1981, 66). If a single conclusion in a constellation of judgments turns out to be weak or false, that will affect the strength of nearby and subsequent conclusions. To the degree that these nearby and subsequent conclusions rely on other conclusions, however, these conclusions will have greater strength than the weak or false conclusion. And so a set of independent supports for the same conclusion produces a far more secure conclusion, a conclusion in which there should be far greater confidence.

Trout (1993) systematizes Wimsatt's efforts by setting three criteria for a robust conclusion: "(1) reproducible by similar methods, (2) detectable by diverse means, and (3) able to survive theoretical integration" (1993, 1). Of these, reproducibility (criterion #1) should be understood as the preliminary requirement of robustness. If a conclusion cannot be reproduced, attempting to detect it through diverse means (criterion #2) may function primarily as a waste of time and resources. Additionally, integrating nonreproducible conclusions into a theory (criterion #3) would not only waste time and resources, but may also challenge unrelated (but true) conclusions supported by the same theoretical perspective.

Reproducibility's preliminary role in distinguishing "genuine discoveries from unchallenged fallacies" highlights how the reproducibility problem could be devastating for medical and psychological sciences. The failure to reproduce the conclusions of medical and psychological science may allow unchallenged fallacies to develop whole lines of inquiry. Take for example the "ego-depletion" line of inquiry in psychological science. For nearly two decades, psychological scientists have been exploring the limits of our reservoirs of self-control based on a model of ego depletion. The theoretical basis for the research is as follows: the more self-control an individual exerts at one moment, the less they have to expend at moments that immediately follow. This line of inquiry has been extensively pursued, with the ground-breaking article (Baumeister et al. 1998) collecting more than 3,000 citations. Even though limits on the ego-depletion perspective have been acknowledged (Hagger, Wood, Stiff, et al. 2010) and explored for more than a decade, it seems that even this work may have been subject to publication bias (Carter and McCullough 2013). Further, a more recent study has shown no evidence of ego depletion at all (Lurguin et al. 2016). Finally, an attempt to replicate the original study, almost 20 years later, has failed to replicate the original results (Hagger and Chatzisarantis 2016). This opens up the possibility that an entire line of inquiry began with an unreproducible study, casting doubt on the entire ego-depletion line of inquiry.

The ego-depletion line of inquiry emphasizes an important point about reproducibility and robustness. Studies that run counter to existing, robustly supported perspectives or that suggest novel and surprising conclusions outside of the domain of current support should be reproduced sooner rather than later. Lacking existing theoretical support, reproducibility (Trout's criterion #1) becomes a key first step in determining the value of a surprising conclusion. It is this that makes the reproducibility problem in younger and/or expanding areas of inquiry significant. A novel study should aim for reproduction first.

After reproduction, however, alternative studies to establish (rather than assume) the fundamental conclusions of the initial study should be pursued (Trout's criterion #2). Indeed, these should be sought in order to rule out the possibility of an isolated effect. Importantly, the fundamental doubts about the ego-depletion line of inquiry may have been discovered earlier if later published studies had not assumed the effect, but instead sought to establish the effect independently (Carter and McCullough 2013). These sorts of publications will, of course, be less enticing and sensational than novelty, but the endless pursuit of minimally supported novelty threatens to produce a house of cards of conclusions.

ROBUSTNESS AND THE EPISTEMIC HUMILITY TAXONOMY

In the initial articulation of the Epistemic Humility Taxonomy, the Defined Domain referred to those applications of medical science that warrant high degrees of confidence. Because this description focused on the contrast between intuitive judgment and medical science, the conclusions of a single research study might have been understood to qualify for the Defined Domain. While a single study may be an improvement on intuitive judgment, it includes a danger of undue confidence. As the ego depletion example demonstrates, if a conclusion has neither been reproduced nor independently established, a similar risk of undue confidence in a conclusion or its application follows and may even lead to less calibrated judgment than intuitive judgment alone. To provide well-calibrated increases in practitioner confidence, a conclusion of medical science must be, at a minimum, reproducible. To justify high degrees of confidence in a practitioner's recommendation, they must be robustly supported.

The fact that single-study conclusions do not meet minimal criteria for high degrees of confidence (as required by the Defined Domain) does not undermine the basic structure of the Epistemic Humility Taxonomy. The goal remains to provide strategies for calibrating the confidence in practitioner recommendations. To achieve this goal, the use of the taxonomy should be reconciled with the requirements of robustness.

One way to conceptualize the Epistemic Humility Taxonomy is less as a series of levels that support for a recommendation moves through and more as a three-dimensional map on which the support for recommendations can be found. The map centers on the Defined Domain, where clusters of robust support for medical science's conclusions lead to recommendations that warrant high degrees of confidence. As the support for a particular recommendation draws on distinct and diverse supports, the recommendation's support comes from within the Defined Domain. The more monolithic or singular the support for the recommendation, the closer the support moves to the Ragged Edge and beyond. As a recommendation depends less directly on clusters of supporting studies, as the support is less robust, the recommendation has less warranted confidence and it moves beyond the Ragged Edge.

For practitioners who are left to make sense of the published literature, this suggests a skepticism about novel and surprising published conclusions of medical science or novel applications of existing treatments. They should attenuate or eliminate their confidence in a conclusion until it has been reproduced and other robust support has been produced. As a short-term measure, simple caution about any published conclusion is necessary. Long-term attentiveness to follow-up studies that reproduce or independently establish novel conclusions is needed. In the meantime, low levels of confidence are appropriate.

For example, look to the patterns of practice for a hypothetical practitioner. When medical science does not provide clear directives, these patterns may be based on her experiences more than anything else. In such cases, she likely recommends a particular treatment based on her intuitive judgment applying the general conclusions of medical science to a particular kind of patient. Even if a patient matches *exactly* the characteristics of previous patients to whom this practitioner has made this recommendation, the recommendation is based on a narrow set (her patients) and subject to a limited analysis (her estimate of the effectiveness of the treatments with a limited, if any, feedback loop). Accordingly, this recommendation is beyond the Ragged Edge of Defined Domain. This, of course, does not mean that her recommendation will turn out to be wrong or that there is some alternative decision-making strategy she should employ—her intuitive judgment may be the best available strategy. It simply means that the recommendation lacks robust support and so the confidence in such a recommendation should be recognized as limited. Imagine now that a recently published study supports her previous recommendations. Imagine also that she verifies that the study is well organized and well executed, and the peer review was rigorous in its evaluation.

Her confidence in the recommendation should remain low. While a single study provides a piece of evidence to support her existing patterns of practice, the reproducibility problem demonstrates that this evidence is weak. The study's conclusion may turn out to be false, and so its effect on her confidence should be nominal or negligible.

Importantly, this kind of support for existing patterns of practice can be particularly pernicious because practitioners in this situation will be particularly predisposed to put too much weight on a single study. Throughout, overconfidence has been a focus of attention, but in this case, two other cognitive biases may exacerbate tendencies toward overconfidence: the status quo bias and the confirmation bias.

The status quo bias refers to decision-makers' tendency to favor current structures, activities, or relationships simply because they are the ones in place. The hypothetical practitioner will be inclined to favor a study that maintains the status quo of her patterns of practice (Samuelson and Zeckhauser 1988). The confirmation bias refers to the unconscious cognitive process by which individuals sort information according to how it relates to their existing views (Nickerson 1998). This sorting serves to support existing views: the sorting process underemphasizes or dismisses information that challenges existing views and overemphasizes information that supports existing views. Both the status quo and confirmation bias predict that this hypothetical practitioner will put too much emphasis on a single study if it supports or confirms her existing beliefs about best patterns of practice. In turn, the status quo and confirmation bias may increase the likelihood of her overconfidence. These concerns about the status quo bias, the confirmation bias, and overconfidence assume, of course, that there is robust support from psychological science that these biases exist and that medical practitioners are subject to them.

RETURNING TO PSYCHOLOGICAL SCIENCE

In the most general terms, there is robust support for the effects of cognitive biases on human judgment. Studies in controlled (Tversky and Kahneman 1974) and uncontrolled (Salovey and Williams-Piehota 2004) environments have demonstrated the biases and their effects on judgment. Additionally, researchers approaching the question of the cognitive limits of human judgment from divergent perspectives (e.g., Kahneman and Tversky [1974] vs. Gigerenzer [1991]) have demonstrated the biases and their effects. The empirical record of psychological science supports a high degree of confidence in the *general* conclusion that cognitive biases have an unconscious effect on decision-making.

The general theoretical conclusions of psychological science should be demonstrated in specific contexts if they are to produce recommendations held with high degrees of confidence. For example, return again to the BDT study. Based on the conclusion of the BDT study, it might seem reasonable to conclude that overconfidence in treatment recommendations is an issue of serious concern within the domain of oncological science. Indeed, for those familiar with the cognitive biases, that oncology practitioners would be overconfident is unsurprising. And yet there are reasons to be less confident in this specific conclusion of the BDT study. On the one hand, this study matches with the general conclusions of psychological science that naive decision-makers and expert decision-makers are both subject to overconfidence (Henrion and Fischhoff 1986; Koehler, Brenner, and Griffin 2002). On the other hand, the BDT study has produced a single-study conclusion. As the reproducibility problem has made evident, a single-study's conclusion may not be reproducible. Indeed, there is no published record of the BDT study's reproduction. In the nearest approximation of this conclusion, Meyer et al. (2013) provide additional evidence of physician overconfidence. This study is not a direct reproduction of the BDT study, so it provides support from an independent perspective. However, this support is complicated in two ways. First, Meyer et al. (2013) focus solely on physicians and not all practitioners. Second, they demonstrate both underconfidence in easy cases and overconfidence in hard cases. This second point matches with other, non-medical practitioner–related research, which has shown overconfidence to be context specific. That is, the manifestations of overconfidence are varied. For routine tasks (e.g., driving a car), decision-makers overestimate their ability, but for unusual tasks (e.g., riding a unicycle), they underestimate it (Moore and Healy 2008). Taking the BDT study and the Meyer et al. study in concert, the confident conclusion is that more research is needed to provide a clear conclusion with robust support.

The BDT study provides a single-study conclusion about overconfidence under the specific conditions of oncological practice. This is important work. But the reproducibility problem in psychological science, combined with only complicated support and the complexities of the general conclusions of psychological science, limits the justifiable confidence in this conclusion. And this problem is not specific to the question of overconfidence among practitioners. Despite the robust support in psychological science for the general effects of cognitive bias on judgment, the applications to medical practice are too often at the Ragged Edge or beyond. Specifically, psychological science provides few robustly supported clear directives for improving judgment in medical practice.

I have suggested previously (Schwab 2008b) a number of areas where clear directives from psychological science might be explored to improve both practitioner and patient decisions. If anything, the reproducibility problem makes the need for this research that much more pressing.

NOTES

1. There is some confusion produced in the literature by the sometimes interchangeable uses of "reproduction"/"reproducibility" and "replication"/"replicability." I will stick to "reproduction"/"reproducibility" throughout for two reasons. First, "replication" can sometimes refer to work done by the original researchers before publishing the results. The concern in this chapter is with production of similar results by others. Second, "replication" also refers to a biological process, making it a tidier distinction to use "reproducibility."

2. I favor the term "practitioner" over "physician" as a term inclusive of all the medical professionals making judgments about patient care.

3. Throughout I refer to "medical science's conclusions" and "psychological science's conclusions." These refer to the published conclusions. While there may be conclusions of medical and psychological science that remain unpublished, these conclusions will remain unavailable to medical practitioners and so are beyond this chapter's scope.

4. Of course, many of medical science's conclusions will fall along a spectrum between these extremes. And when they do, the diversity of recommendations and the confidence in those recommendations should shift accordingly. If the BDT study turns out to be correct, overconfidence will be a problem in these spaces as well.

5. I will return to reproducibility and psychological science at the end of this chapter.

6. At present, the Reproducibility Project: Cancer Biology is attempting to reproduce 50 high-impact studies (http://elifesciences.org/content/3/e04333v1).

7. What remains unclear is the extent of the reproducibility problem in medical and psychological science. The more isolated the problems of reproducibility are from each other, the more the challenge remains limited to the confidence in particular conclusions, and not to any general theoretical perspective. If, however, there are sets of interrelated conclusions that are not reproducible, this may require substantive revision of the theoretical perspectives to which these conclusions have given rise. Accordingly, a point of emphasis in the analysis of the reproducibility problem for both medical and psychological science hinges on the mapping of reproducibility problems.

The least amount of disruption to medical and psychological science will be if the reproducibility problems are isolated from one another. In this case, the theoretical perspectives will not be challenged, but only particular conclusions. From the perspective of the psychological and medical scientist, this would be the best outcome from the reproducibility problem. For practitioners and their patients, however, this will be cold comfort. They are seeking a treatment recommendation for a particular condition in which they can have high degrees of confidence. Or at least one in which they know how confident they should be.

8. This is not to say that the Epistemic Humility Taxonomy is incompatible with other epistemological perspectives, but that reliabilism is the favored epistemological perspective.

9. As we'll see in the next section, the most robust conclusions will not only be reliable in this narrow way, but will also be integrated into the background theory.

10. This point is perhaps too poignant in an era of managed care.

11. Importantly, Wimsatt analyzes robustness as a feature of any kind of judgment. His chapter includes references to the robustness of our experience of seeing a table, for example. The discussion here will be limited to discussions of robustness as related to scientific conclusions. This is not meant to indicate that robustness is limited in this way in general, but that its significance for this project is limited to its implications for specific conclusions drawn in medical and psychological science.

REFERENCES

Baumann, Andrea O., Raisa B. Deber, and Gail G. Thompson. 1991. "Overconfidence Among Physicians and Nurses: The 'Micro-Certainty, Macro-Uncertainty' Phenomenon." *Social Science and Medicine* 32: 167–174.

Baumeister, Roy F., Ellen Bratslavsky, Mark Muraven, and Dianne M. Tice. 1998. "Ego Depletion: Is the Active Self a Limited Resource?" *Journal of Personality and Social Psychology* 74(5): 1252–1265. http://dx.doi.org/10.1037/0022-3514.74.5.1252

Begley, C. Glenn, and Lee M. Ellis. 2012. "Drug Development: Raise Standards for Preclinical Cancer Research." *Nature* 483: 531–533.

Carter, Evan C., and Michael E. McCullough. 2013. "Is Ego Depletion Too Incredible? Evidence for the Overestimation of the Depletion Effect." *Behavioral and Brain Sciences* 36 (6): 683–684.

Doshi, Peter, Steven N. Goodman, and John P. A. Ioannidis. 2013. "Raw Data from Clinical Trials: Within Reach?" *Trends in Pharmacological Sciences* 34 (12): 645–647.

Easterbrook Philippa J., Jesse A. Berlin, Ramana Gopalan, and David R. Matthews. 1991. "Publication Bias in Clinical Research." *Lancet* 337: 867–872.

Gigerenzer, Gerd. 1991. "How to Make Cognitive Illusions Disappear: Beyond 'Heuristics and Biases'." In *European Review of Social Psychology*, Vol. 2, edited by Wolfgang Stroebe and Miles Hewstone, 83–115). Chichester, UK: Wiley.

Hagger, Martin S., and Nikos L. D. Chatzisarantis. 2016. "A Multi-Lab Pre-Registered Replication of the Ego Depletion Effect." *Perspectives on Psychological Science* 11 (4): 546–573. http://www.psychologicalscience.org/index.php/publications/rrr-the-ego-depletion-paradigm

Hagger, Martin S., Chantelle Wood, Chris Stiff, and Nikos L. D. Chatzisarantis. 2010. "Ego Depletion and the Strength Model of Self-Control: A Meta-Analysis." *Psychological Bulletin* 136 (4): 495–525. http://dx.doi.org/10.1037/a0019486

Henrion, Max, and Baruch Fischhoff. 1986. "Assessing Uncertainty in Physical Constants." *American Journal of Physics* 54: 791–798.

Hirt, Edward R., and Keith D. Markman. 1995. "Multiple Explanation: A Consider-An-Alternative Strategy for Debiasing Judgments." *Journal of Personality and Social Psychology* 69: 1069–1086.

Hirt, Edward R., Frank R. Kardes, and Keith D. Markman. 2004. "Activating a Mental Simulation Mind-Set through Generation of Alternatives: Implications for Debiasing in Related and Unrelated Domains." *Journal of Experimental Social Psychology* 40: 374–383.

Ioannidis, John P. A. 2005. "Why Most Published Research Findings are False." *PLOS Med.* 2: e124.

Ioannidis, John P. A. 2012. "Why Science Is Not Necessarily Self-Correcting." *Perspectives on Psychological Science* 7 (6): 645–654.

Koehler, Derek J. 1991. "Explanation, Imagination, and Confidence in Judgment." *Psychological Bulletin* 110: 499–519.

Koehler, Derek J., Lyle Brenner, and Dale Griffin. 2002. "The Calibration of Expert Judgment: Heuristics and Biases Beyond the Laboratory." In *Heuristics and Biases: The Psychology of Human Judgment*, edited by Griffin Gilovich and Daniel Kahneman, 686–715. New York: Cambridge University Press.

Koriat, Asher, Sarah Lichtenstein, and Baruch Fischoff. 1980. "Reasons for Confidence." *Journal of Experimental Psychology* 6: 107–118.

Lurquin, John H., Laura E. Michaelson, Jane E. Barker, Daniel E. Gustavson, Claudia C. von Bastian, Nicholas P. Carruth, and Akira Miyake. 2016. "No Evidence of the Ego-Depletion Effect across Task Characteristics and Individual Differences: A Pre-Registered Study." *PLOS One*. Accessed June 3, 2016. dx.doi.org/10.1371/journal.pone.0147770

Meyer, Ashley N. D., Velma L. Payn, Derek W. Meeks, Radha Rao, and Hardeep Singh. 2013. "Physicians' Diagnostic Accuracy, Confidence, and Resource Requests: A Vignette Study." *Journal of the American Medical Association* 173 (21): 1952–1959.

Moore, Don A., and Paul J. Healy. 2008. "The Trouble with Overconfidence." *Psychological Review* 115 (2): 502.

Nickerson, Raymond S. 1998. "Confirmation Bias: A Ubiquitous Phenomenon in Many Guises." *Review of General Psychology* 2 (2): 175–220.

Open Science Collaboration. 2015. "Estimating the Reproducibility of Psychological Science." *Science* 349 (6251): aac4716–1–8. doi: 10.1126/science.aac4716

Pashler, Harold, and Christine R. Harris. 2012. "Is the Replicability Crisis Overblown? Three Arguments Examined." *Perspectives on Psychological Science.* 7 (6): 531–536.

Prinz, Florian., Thomas Schlange, and Khusru Asadullah. 2011. "Believe It or Not: How Much Can We Rely on Published Data on Potential Drug Targets?" *Nature Reviews Drug Discovery* 10: 712–713. doi:10.1038/nrd3439-c1

Salovey, Peter, and Pamela Williams-Piehota. 2004. "Field Experiments in Social Psychology: Message Framing and the Promotion of Health Protective Behaviors." *American Behavioral Scientists* 47: 488–505. doi:10.1177/0002764203259293

Samuelson, William, and Richard Zeckhauser. 1988. "Status Quo Bias in Decision Making." *Journal of Risk and Uncertainty* 1: 7–59.

Schwab, Abraham P. 2008a. "Epistemic Trust, Epistemic Responsibility, and Medical Practice." *Journal of Medicine and Philosophy* 33: 302–320.
Schwab, Abraham P. 2008b. "Putting Cognitive Psychology to Work: Improving Decision-Making in the Medical Encounter." *Social Science and Medicine* 67 (11): 1861–1869.
Schwab, Abraham P. 2012. "Epistemic Humility and Medical Practice: Translating Epistemic Categories into Ethical Obligations." *Journal of Medicine and Philosophy* 37: 28–48.
Simon, Richard, and Sameek Roychowdhury. 2013. "Implementing Personalized Cancer Genomics in Clinical Trials." *Nature Reviews Drug Discovery* 12: 358–369. doi:10.1038/nrd3979
Trout, J. D. 1993. "Robustness and Integrative Survival in Significance Testing: The World's Contribution to Rationality." *British Journal of Philosophy of Science* 44: 1–15.
Tversky, Amos, and Daniel Kahneman. 1974. "Judgment under Uncertainty: Heuristics and Biases." *Science* 185 (4157): 1124–1131.
Wimsatt, William C. 1981. "Robustness, Reliability, and Overdetermination." In *Scientific Inquiry in the Social Sciences*, edited by Marilynn B. Brewer and Barry E. Collins, 123–162. San Francisco: Jossey-Bass.

Chapter 5

Reframing a Model

The Benefits and Challenges of Service User Involvement in Mental Health Research

Tania Gergel and Thomas Kabir

"Nothing about us without us," a slogan adopted more generally within the broader disability rights movement (Charlton 1998), has become a key principle behind the promotion and incorporation of service user involvement in mental health research.[1] Both within service user groups and mental health services themselves, there is a growing sense that proper involvement of those with "lived experience" of mental health conditions in research into such conditions will have major benefits on an epistemological, metaphysical, and ethical level, insofar as service users themselves are being recognized as an invaluable source of expertise on how best to investigate mental health conditions, what constitutes the true nature of such conditions, and what constitutes the most appropriate and just way to treat those who suffer from such conditions. Not only is such an approach viewed as a way of relieving some of the disempowerment and stigmatization that has afflicted mental health service users, but at the same time as a way of improving the quality of research itself.

After briefly outlining the history and background of service user involvement in mental health, we will explore the potential benefits and consider some specific examples to see whether and to what extent such benefits have been realized within existing research. Finally, we will consider the challenges that still face involvement and the ways in which such challenges might be overcome.

THE HISTORY OF MENTAL HEALTH TREATMENT
AND RESEARCH: INSIGHT, CONTROVERSY,
AND THE "DEFICIT" MODEL

The demand for UI [user involvement] has been particularly significant within mental health services, not only in the UK but throughout the developed world …. The stigma of mental illness, the fight for recognition of intellectual capacity, and the historical use of physical and pharmaceutical control and restraint unite psychiatric patients and "survivors" in a cause which encompasses, but is not limited to, the reform of services. (Rutter et al. 2004)

The most severe episodes of mental disorder are often characterized by what is termed "a loss of insight," meaning that the individual who is experiencing the illness is very likely, at times, to believe that they are not, in fact, unwell or to dispute key aspects of the psychiatric assessment and prescribed treatment. Thus, for mental health service users, "lived experience" of illness frequently involves being diagnosed and treated, often involuntarily at the time, for an illness whose diagnosis and treatment they themselves may well question. This makes the lived experience of mental health users bound up in a particular set of complexities that are not found in relation to other health conditions and that, very likely, go much of the way to explaining the particular prominence of user involvement within the context of mental health.[2]

For mental health service users, therefore, dispute is frequently fundamental to lived experience, and clinical practice within mental health is characterized by a challenging tension. On the one hand, there is the importance of avoiding excessive paternalism, the need to respect the individual's opinion and to protect their right to freedom of choice. On the other, some degree of paternalistic intervention is, all too often, unavoidable, if the clinical team is to protect the individual's right to health and their right to be protected from the damage that may ensue for themselves and others as the result of failure to treat.

Accordingly, in many jurisdictions, including the UK, there is special statutory provision for involuntary detention and treatment of those with severe mental health conditions based on "status," meaning the presence of a recognized mental disorder, and "risk," meaning that, were the affected individual to remain at liberty and untreated, they are considered to present a serious risk either to themselves or to others (Dawson and Szmukler 2006). Mental disorder is therefore the only context within the law in which there is a provision for preemptive detention on the grounds of risk of harm, as opposed to after harm has already been committed, which is frequently employed.[3]

The very nature of mental disorder means that questioning the lived experience of the mental health service user, while he or she is unwell, is a fundamental and understandable aspect of clinical practice. In addition, however, the complexities surrounding mental health are heightened by continuing controversy over diagnosis and diagnostic models of mental disorder, so that, from an ontological perspective, psychiatry remains perhaps the most uncertain of all medical disciplines (see, e.g., Bolton 2009). The classifications of mental disorder within the two most established medical diagnostic manuals, the *Diagnostic and Statistical Manual of Mental Disorders* (*DSM*) and the *International Classification of Diseases* (*ICD*), not only differ from each other in certain key details, but are both subject to frequent change and revision, and open to challenge from other models such as the psychological and sociological. Furthermore, there is evidence that the diagnostic criteria set out by manuals such as the *ICD* are unstable and poorly applied in practice (e.g., Baca-Garcia et al. 2007).

The evidence base for the success of many conventional psychiatric treatments is relatively limited when compared to treatments for other conditions, while there is often insufficient understanding and consideration of the nature, severity, and impact of side effects. Corstens et al., for example, reflect on the limitations of conventional treatments for the phenomenon of "hearing voices":

> Our current methods of supporting voice-hearers within psychiatry are often based on limited evidence. The evidence for the long-term effectiveness of pharmacotherapy, the dominant treatment for psychosis, is not well substantiated and its more general hazards and evidential limitations for voices specifically are insufficiently acknowledged. (Corstens et al. 2014, 291)

Overall, with its fluctuating nature and lack of biomarkers, dissatisfaction with diagnosis and treatment of mental disorder is common, as well as poor adherence to medical treatment when prescribed.

A degree of paternalism and controversy over diagnosis and treatment of mental disorder can be explained by issues such as loss of insight, the lack of physiological markers, and uncertainties surrounding the effectiveness of available treatments. Nevertheless, historically, and even today, people who experience mental health conditions have been subject to excessive degrees of paternalism, doubt, and abuse. A stigmatizing "deficit" model leads to a view that mental illness renders an individual, even when well, deficient in ordinary human mental abilities, such as rationality, control, and decision-making (see, e.g., Gergel 2014).

This model has manifested itself in various ways, and those with mental health conditions have fallen victim to what are often seen as the two key and interrelated types of injustice affecting those with disabilities: disrespect and distributive inequity. "Disrespect" refers to the unequal treatment of those for whom there is no legitimate basis for judging them as inferior in any respect; "distributive inequity" is the use of fallacious and unjust assessments of inequality as a justification for unequal distribution of resources to those considered to be inferior.

In economic terms, proportionately low resources are devoted to mental health conditions and their consequences, when compared to the disease burden.[4] Historically, there has also been a widespread history of abuse of those with mental disorder, as well as manipulating a diagnosis in order, for example, to serve political ends. Examples include the Nazi persecution of the mentally ill and the systematic political abuse of psychiatry in the Soviet Union and communist China, with political opposition or dissent diagnosed as a psychiatric condition and resulting in detention and treatment within psychiatric institutions (see, e.g., Bonnie 2002).

Understandably, the particular complexities of mental health described above often lead to difficulties within the relationship between service user and clinician and may be characterized by phenomena such as resentment, disempowerment, and loss of trust, which can, in turn, have negative effects on conventional research. Service users may well be discouraged from taking part in research if, for example, they perceive it as reflecting questionable values, opinions, and illness models or if they have concerns about respect for their rights and opinions during the research process itself. Service users are also likely to question the veracity, relevance, and usefulness of research outcomes, if it is felt that they are informed by a model of psychiatric understanding that does not reflect their own experiences, opinions, or priorities.

THE DEVELOPMENT AND BENEFITS OF SERVICE USER INVOLVEMENT IN RESEARCH IN MENTAL HEALTH

One of the driving principles behind service user involvement in research in general is that involvement will yield fairer and more accurate research into mental health conditions, which is simultaneously more truly reflective both of service user best interests and of the nature of the conditions themselves. Although it began as a breakaway movement, running alongside conventional biomedical research, an increased recognition of the benefits that more thorough service user involvement yields for research means that it is moving increasingly toward the mainstream of mental health research.

Although user involvement in health research is a relatively recent innovation, within mental health it has a relatively long history when compared with many other areas. Some of the earliest published examples of involvement in mental health research date back to the 1970s. For example, in 1978 Leonard Roy Frank published his research into ECT after having the procedure himself 35 times in the USA.[5] In the 1980s and 1990s, involvement in mental health research began to take root in the UK, and in 1996 the UK Department of Health funded a group now known as INVOLVE (www.invo.org.uk) to "support active public involvement in NHS, public health and social care research. It is one of the few government-funded programmes of its kind in the world." While INVOLVE did much work to define and support active involvement in research, there were also a growing number of mental health service users becoming researchers in their own right.[6]

The UK's largest funder of health research, the National Institute of Health Research (NIHR), went on to make involvement an expectation within the research that it funds: "We expect the active involvement of members of the public in the research that we fund. We recognise that the nature and extent of active public involvement is likely to vary depending on the context of each study or award."[7]

This expectation has resulted in a large increase in the number of research projects involving service users or carers, and the UK is now considered to be one of the leaders in involvement in health research.[8]

The principle reported benefits of user involvement in health research are that it improves the quality and relevance of the research, while also ensuring that research is ethically sound.[9] User involvees can help to ensure that chosen research topics are important to other service users and that resources are used efficiently. They may be able to identity new potential areas for research, enlarge upon those chosen by health or social care professionals working in isolation, and help with inclusivity. Service users can also monitor and help to reshape or clarify research, so that the language and content of information provided to participants is more appropriate and accessible and the research methods are acceptable and sensitive to the situations of potential research participants.[10]

The term "involvement" itself designates active involvement by service users in the planning, development, and running of research projects, as opposed to simply participating as passive research subjects in clinical trials or other research. INVOLVE defines "involvement" as

> research being carried out "with" or "by" members of the public rather than "to," "about" or "for" them. This includes, for example, working with research funders to prioritise research, offering advice as members of a project steering group, commenting on and developing research materials, undertaking interviews with research participants. (INVOLVE 2012)

Central to involvement, therefore, is the switch from a passive to an active role, but also the principle of "continuity," in which involvement occurs throughout the research process, as opposed to being limited to particular stages or aspects. Involvement is found within all the following stages of research, often featuring within multiple stages of the same research project: planning and study design, recruitment, data collection, analysis, dissemination of results, and education. In general, it is believed that involvement will help to reverse some of the power imbalance that has occurred within mental health care and research, allowing service users to have more control and influence over research agendas and methods. At the same time, involvement is viewed as improving the accuracy and productivity of research, leading to a combination of more successful outcomes, but also benefits for the individual service users themselves from the process of involvement.

In very broad terms, involvement can be divided into three key areas (Szmukler et al. 2011):

- Consultation
- Collaboration
- Control

However, it is important that these terms should not be seen to denote a hierarchy: "Over time it has become clear that in practice researcher can use a combination of these three and it is more helpful to describe them as approaches rather than levels" (INVOLVE 2012).

Before continuing to explore more specifically the benefits of involvement for the usual expected outcomes of research, we note that it is also the case that all three of these elements can be seen to confer substantial moral benefits. By making "consultation" with service users a routine element of the planning stage of research, viewed as beneficial for research outcomes, involvement allocates epistemological value to the knowledge that service users themselves have gained through firsthand experience of their conditions and goes some way to reverse the epistemological injustice, of which service users have been victims, in which their testimony has been disregarded as lacking in value or credibility by virtue of the presence of mental disorder (Sanati and Kyratsous 2015). It is hoped that the opinions gained through consultation with service users will have a substantive impact on factors such as the relevance of the question, the design of the project, and the appropriateness of the outcome measures.

Attributing epistemological value to the opinions of service users is increased through "collaboration," a term that here denotes involvement of the service user as an active partner within the research process over the course of the project. In a sense this reverses the traditional power imbalance involved in the deficit model, insofar as the service user is elevated from the role of passive subject to active participant, thereby creating a sense of equality through a "commitment from the research team that control over the project will be shared to a greater or lesser extent" (Szmukler et al. 2011). Finally, there is "control," in which "decision-making concerning the objectives, design, outcome measures, and so on, lie with service users rather than professional researchers," a level of involvement that takes this "emancipatory" element even further.

As well as going some way to address the sense of disempowerment often experienced by mental health service users, involvement can bring further benefits to service uses. A 2016 study conducted by Ashcroft et al. found that service users reported a mixture of psychological, social and intellectual benefits, as well as an improved relationship with illness and crisis, from becoming involved in research (Ashcroft et al. 2016).

There is then a strong moral case to be made for increased involvement in terms of the empowerment and other benefits it brings to service users. However, at the same time, proponents of service user involvement argue that it improves the quality and productivity of the research process and outcomes themselves, pointing to the growing body of evidence to suggest the ways in which involvement brings benefits to a diverse range of research outcomes, including the relevance of research findings to patients and carers, identification of research questions and shaping or reshaping study design, recruitment, data analysis, and information gathering (Szmukler et al. 2011). Moreover, research can be a forum within which service users can challenge the very model of illness that is being investigated and where service users can play an active role in dissemination and education.

In many ways, service user involvement in mental health research can be seen as an augmentation of the knowledge model, in both how it is conducted and the outcomes it achieves. No longer is the service user the passive "data" for scientific research, analyzed by others to achieve a target of improved treatment. A growing body of evidence is coming to show that, with involvement, the active participation of those whose knowledge is generated through "lived-experience" is an invaluable element of knowledge generation. Moreover, with the varied improvements that involvement can offer to service users in terms of managing their mental health condition and general well-being, research or knowledge generation and treatment are conflated in a way that goes far beyond the model of a clinical trial.

TRANSFORMING A MODEL: SOME SPECIFIC EXAMPLES

Thus, involvement can be seen to offer a tripartite transformation of mental health research that cuts across the ethics and metaphysics, as well as the epistemology of psychiatry. This can, perhaps, be best explained by the consideration of some specific examples that show how such a transformation is manifested both across the range of the different components of research and across a range of different mental health conditions and phenomena. It is important to point out that the examples below are simply a small sample of existing research.

For example, as well as improving our knowledge of existing treatments or helping to improve the quality of life for those who experience recognized conditions, involvement might actually contribute to creating a different understanding of what constitutes pathology. A long-standing and influential service user movement, which began in the 1980s and has generated a broad range of research and other activities, is the "Hearing Voices Movement (HVM)," the "main tenet" of which is "the notion that hearing voices is a meaningful human experience" (Corstens et al. 2014a, 45), as opposed to being merely a symptom of mental disorder to be decreased or eradicated wherever possible through psychiatric treatment. While HVM began very much as a breakaway movement, in opposition to conventional research and the medical establishment, in recent years there has been more research, both within the movement itself and in growing collaborations with clinicians. The engagement of HVM with research extends to all aspects of research and presents a very clear example of the tripartite transformation of the knowledge model: "The HVM approach necessitates full participation of experts by experience at all stages. This includes setting questions and developing methodologies" (ibid.).

HVM challenges the fundamental understanding of the nature of hearing voices and the classification of such phenomena purely as auditory delusions, understood solely within an illness paradigm and biomedical frame of reference. Central to the HVM movement and to any research with which it is engaged is a commitment to some level of ontological validity and value to "voices" and a rejection of the view that they are, fundamentally, pathological phenomena. Along with this comes a substantive reconsideration of the mechanisms of knowledge acquisition, the sources of expertise, and the aims and intended outcomes of furthering knowledge and understanding in this area. The ramifications of such shifts go far beyond any narrow academic or clinical sphere. HVM has been instrumental in combating the stigmatization and social isolation of voice hearers and in helping them to become more integrated and accepted members of society. Research within HVM is also deeply interdisciplinary and is a prime example of where, for example, a philosophical and humanities-based approach to what is generally seen as a clinical phenomenon can serve to open up new knowledge-gathering approaches (Fernyhough 2016).

Not only has the HVM movement been involved in the planning and design stage of the research process, involvement has then continued throughout all stages of collaborative research. For example, although trials continue to use standard outcome measures, which focus primarily on overall reduction of voice hearing or aspects of the voices, there has been recent development of service-user-informed measures of "the subjective experience of psychosis, CBT for psychosis outcome, and personal recovery, while service users themselves have benefited from taking part in research and have also started to provide training to academics and mental health professionals" (Corstens et al. 2014). For those who hear voices, HVM, with its research and other activities such as peer support, is opening up a fresh way of understanding and managing such phenomena, which goes beyond conventional medical approaches and offers new hope and a significantly different approach.

There are numerous other examples of projects where service users have made a major contribution to study design and planning. While dominant academic practices prioritize the role of the professional researcher in devising the aims and methods and carrying out collection and analysis of data, projects with direct involvement of service users in these activities offer a new epistemological model based on the value of knowledge through lived experience. One such project was the development of a new way for measuring patient views on inpatient care (VOICE, Views on Inpatient Care), which was developed using a "participatory methodology to maximise the opportunity for service user involvement," which aimed to redefine the outcome measures to make them more consistent with service users' priorities (Evans et al. 2012). Similarly, a slightly later project uses a participatory methodology, which uses user-generated measures of the impact of ward design, to capture service user and staff perceptions. As the authors comment, UK "government reports increasingly emphasise the importance of ward design for patient well-being and recovery" (Csipke et al. 2016), and their study revealed that a measure designed through collaboration with service users is an important resource in the process of evaluating inpatient facilities.

In another project, service user involvement was fundamental to the design of a new intervention aimed to help those with bipolar disorder. A trial carried out across the northwest of England used extensive participatory methods to include those with lived experience of bipolar disorder to devise a method of recovery-focused CBT (cognitive behavioral therapy) for those in the early stages of bipolar disorder. As the authors comment, "This level of engagement of individuals with personal experience of bipolar disorder fits with the model of recovery approaches as being empowering, individualised and grounded in the individual's own priorities and needs" (Jones et al. 2012). The use of participatory methods in both the ward design and bipolar CBT studies are good examples of how a change in the model of knowledge acquisition, with the knowledge and understanding of conditions and treatments acquired through lived experience playing a fundamental role in intervention design, can lead to more effective interventions that yield substantive health benefits to service users.

Yet, involvement with services is not only restricted to those who themselves experience a mental health condition, but also includes carers, and the movement toward involvement has also been active among this group, where it has, in general terms, yielded similar benefits to those outlined with the two studies mentioned above. One 2015 study led by a researcher who is herself a carer used a series of interviews and focus groups, developed in partnership with an advisory group of carers and service users to assess carers' views on the care planning process for people with severe mental illness (Cree et al. 2015). Once again, the team who analyzed the data was a mixture of service user and carer co-applicants working alongside experienced qualitative researchers. Overall, the project was said to have "developed the understanding of the potential role of carers within the care planning process," while it was also "likely that these findings have direct policy and practice relevance." Moreover, the fact that service users and carers themselves also took part in analysis of data presented a clear challenge to conventional practices in which only those with a professional training in data analysis are seen to be able to make a valuable contribution to this element of research.

Another study, led by Gillard at St George's Hospital, University of London, focused specifically on the impact made on the research process by involving service users in the collection and analysis of research data and comparing the results with collection and analysis conducted by conventional university researchers (Gillard 2010). Gillard concludes that the study not only "demonstrates the potential to develop a methodologically robust approach to evaluate empirically the impact of SURs (service user researchers) on research process and findings," but also indicates the potential benefits of involving service users in the research process. In this case "service user researchers" refer to service users who were employed as researchers in their own right alongside other members of the study team. While the study was limited by the small number of participants, the findings were nevertheless interesting. The study was a set of interviews asking inpatients detained under the Mental Health Act questions about their experience of formal detention within acute and forensic psychiatric wards. The interviewers were free to ask follow-up questions that seemed to be relevant in addition to the original set questions, and it was these that were analyzed. While SURs were proportionately more likely to ask questions about "experiences and feelings," university researchers were proportionately more likely to ask questions relating to medical treatments, seeking to understand experiences from a behavioral perspective and relating to the extent to which the interviewees felt any sense of agency or control during their detention. The Gillard study therefore suggests that qualitative interviews when conducted and analyzed by a collaborative team including service users may result in a broader range of results, reflecting different priorities, due to the different training, experiences, and priorities of those collecting the data. In this particular example, it seems that the service user researchers have a different understanding of what constitute the most significant elements of the detention experience from those aspects valued by professional researchers.

Furthermore, although interviewees could not definitely be sure as to whether it affected the answers that they gave, all those who were interviewed by a service user described it as a positive experience, in which they felt "more comfortable" and "some stated that they found it personally encouraging meeting a service user working as a researcher." This suggests that, from an ethical perspective, involving service users in data collection during mental health research may go some way to counter the feelings of mistrust or disempowerment experienced by service users in relation to the medical establishment. Similarly, one of the motivations for Svennson and Hansson, who themselves conducted a study in which service users were trained to be interviewers for fellow service users regarding satisfaction with inpatient and outpatient mental health services, was that in some earlier studies "patients interviewed by user interviewers gave more negative responses about services and lower service satisfaction scores," suggesting perhaps that service users are more comfortable in expressing dissatisfaction to a fellow service user than to a mental health professional (Svensson and Hansson 2016). A 1993–1994 study conducted in Toronto found that "clients gave more extremely negative responses to client interviewers" and, although they point out that the small numbers make any general conclusions difficult, they suggest the possibility that "greater feelings of safety, trust, confidentiality, and privacy may have influenced the interviewees disclosures to other clients," while service users would be worried about expressing disapproval to staff members. They conclude that, in order to attain more valid feedback about service user views, there should be more opportunities for "clients to address issues with other clients in privacy" (Clarke et al. 1999). A 2002 systematic review of "involving users in the delivery and evaluation of mental health services" also reports that "clients reported being less satisfied with services when interviewed by other users of the service in evaluation research" (Simpson and House 2002). The implication here is that service users may feel able to be more open about their experiences with researchers who identify as having suffered for mental health problems themselves. This suggests that, for mental health research to gain the most accurate understanding of the nature of mental disorders and its treatments, and how these might be improved, the inclusion of service users within the research process should be seen as a necessary element.

A 2013 review by Ennis and Wykes shows that patient involvement in mental health research also leads to more successful recruitment for studies and can therefore be associated with study success. The reasons that they suggest for these findings also reflect the difference that lived experience can make to the research process, ranging from practicalities such as making the study information sheets more manageable and appealing "because of vetting by other patients to influencing the project design itself through their own insight into the realities of living with a mental health problem" (Ennis and Wykes 2013). Furthermore, they also suggest that the "principle of patient involvement is in itself appealing" to patients who are considering taking part in research. Although the authors do not give reasons for this suggestion, it is possible, once again, to see this in terms of fostering a sense of trust and empowerment for potential research subjects, through their peers being involved as active researchers.

Finally, along with the various potential benefits to the research outcomes themselves from involving service users in the research process, involvement in research can also have a beneficial impact on the well-being of the service user involvees themselves. For example, a participatory research partnership that involved drug users in all aspects of a study to HIV risk and prevalence among drug users in Ottawa not only led to successful research outcomes, but also had significant benefits for the service users through their involvement in the project in terms of increasing empowerment and providing develop-ment opportunities (Lazarus et al. 2014). Allowing medical students and drug users to work alongside each other "worked to challenge traditional power imbalances between health professionals" and the training helped to "foster relationships and build trust," as well as allowing "providing meaningful lead-ership roles" to drug users. On being a "user and an interviewer," Svensson and Hansson report that "a majority of the interviewers found their commis-sion rewarding and positive" (2016), while Patterson's survey of views of service users involved in mental health research reports that the majority saw service users as "empowered through their active participation in research" (Paterson et al. 2014).

Another motivation and benefit to service users from involvement is that this provides an opportunity to allow challenging life experiences to lead to some positive and productive outcomes. A survey by Ashcroft et al. reports that "the wish to gain personal meaning from the experience of illness and to contribute something useful for others was widely stated" and that "the majority of respondents identified with largely altruistic motivations for get-ting involved in research" (2016). Ashcroft et al. go on to report that "another motivation reported by the majority of respondents was to feel useful (55%). The lowest reported motivations were gaining financial income (9%) and improving skills or gaining work experience (29%)."

It is clear from these examples that the involvement of those with lived experience can have a major impact on the planning and design of research projects, methodology, outcomes, concepts of health and illness, and even the design of the very interventions that are under investigation and that this extends to a range of different conditions and aspects of clinical practice. Prioritizing lived experience within the research process not only changes the epistemological model, but can produce new understandings of the ontology of the conditions and experiences being researched, all of which have significant moral benefits, from the increased well-being yielded by more successful research outcomes to the increased empowerment and trust that can be engendered among service users.

ONGOING CHALLENGES FOR SERVICE USER INVOLVEMENT IN RESEARCH AND HOW THESE MIGHT BE OVERCOME

Many of the challenges that still face service user involvement in research could be broadly termed as political or practical.

Political Barriers to Involving Service Users and Carers in Research

The political issues relate to issues such as acceptance, integration, diversity, and conformity of user involvement within mental health research in general. While occasional concerns are still expressed about user involvement being overly radical or politicized and the difficulties of integrating this into the broader research context, a much more widespread concern seems to be the very opposite. This is the worry that, as service user involvement becomes an accepted and even expected part of mainstream mental health research, it will lose sight of the pioneering and transformative agenda that informed the initial projects, as it starts to conform to the more established biomedical and academic models, as a result of increased collaboration. In the words of Turner and Gillard, "Service user involvement in mental health research is no longer confined to small-scale initiatives led by pioneering individuals. Service user involvement in research has mainstreamed and has moved into the university" (2011). Does such mainstreaming mean that its transformative dimension is lost or compromised?

Turner and Gillard discuss how the academic research environment might contribute in certain ways to restraining the authentic voice of the service user and the obstacles "within the academic environment, where prevailing scientific practice can act to sanitise or stifle the service user voice." For example, even though the service user is often brought into a project for the very reason that they may be able to offer an alternative approach or perspective on the research area, there may be a view that good research depends upon the "inter-rater reliability" of the service user's analysis with that produced by the other researchers: "The service user voice was perceived as having validity where it was shown to be in agreement with those more conventional research voices." There was even a suggestion that a "lack of distinctive analysis signified that service user involvement had been a success; that the service user was a good researcher."

Turner and Gillard conclude that the service user voice can still be heard and have influence as long as the research mainstream moves to accommodate the service user, rather than the expectation for adaptation to research methods and agenda lying solely with the service users. Similarly, a study by Rutter et al., which looked at user involvement in planning and delivery of services, concluded that greater openness was needed on the part of the National Health Service (NHS) Trusts,[11] rather than the expectation for conformity to Trust agendas and management practices currently placed on service user involves (Rutter et al. 2004).

In what ways, then, can conventional research frameworks adapt in order better to accommodate service user involvement in a manner that is a genuine reflection of the priorities and opinions of the user involvees and truly reflects the transformative agenda at the heart of the lived experience knowledge model? A key part of this is to convince those within the mainstream research environment that user involvement adds real value to research and is not simply a bureaucratic pressure or expectation, in order to ensure the allocation of essentials such as funding, time, and training. As Corstens et al. write, in relation to the Hearing Voices Movement, truly collaborative research

> requires an investment of resources into training individuals with lived experience to understand research methodology and practice in order to meaningfully contribute to both study design and interpretation. It also requires funders to recognize the value of collaborative research and to consider prioritizing initiatives that demonstrate involvement. (Corstens et al. 2014)

In order for this to be accomplished, it may be necessary for the academic framework itself to change in certain respects in a way that makes room for the contribution of service users to be recognized as bringing value and to be subject to fair assessment. A UK survey of service user involvees reports that a frequent challenge to successful involvement was that, among research professionals, "expertise grounded in lived experience was not really valued and that different types of expertise were not accorded equal status," while service user input "was discounted as non-academic, subject to excessive scrutiny for bias or conversely overvalued rather than being subject to critical review" (Patterson et al. 2013). Concerns are also often raised about tokenistic inclusion of service user endorsement; as one service user reports, "tokenism,

tokenism, tokenism … You are there because the funders have asked to show service user participation" (ibid.; see also Ashcroft et al. 2016).

It is also important to ensure that user involvement becomes truly representative across social and national boundaries, rather than being restricted to certain groups of service users or particular countries where the value of such research has been recognized. For example, Semrau et al. point out the lack of involvement of service users in research with low- and middle-income countries and make various recommendations for how this might be achieved and areas of research, such as study design, where service users could make valuable contributions (Semrau et al. 2016; Corstens et al. 2014). A UK survey of service users' experience of being involved in research conducted by Ashcroft et al. pointed to the need for a greater diversity of composition of the patient and carer involvees: "Respondents recognised the need to better involve minority ethnic groups and people with low income" (Ashcroft et al. 2016). Another factor in ensuring that involvement is truly representative is not simply ensuring that a wide range of service users are included and valued, but also prioritizing "continuity" and ensuring that involvement is not restricted to certain elements of research, but is spread throughout the research process (Callard et al. 2012).

Moreover, if the research framework is truly to be changed to recognize the value and range of lived experience, then research agenda and desired outcomes must also be adapted to accommodate and reflect the views and priorities of service users. From the language used to the outcomes sought, it is important to recognize that shifts may need to take place within the existing clinical conventions. Corstens et al. suggest that "finding truly neutral and descriptive language that is not based within the illness paradigm" might help to "engage people in research who would normally avoid it" (2014).[12] Trivedi and Wykes discuss, for example, service users' rejection of existing clinical outcome measures such as "insight and compliance" and their explanation that "although these outcomes might be very important for clinicians, they were anathema to many users who perceived them as echoing the paternalistic and disempowering authority of psychiatry" (2002). Some service users have attempted to address this issue by constructing outcome measures either themselves or in conjunction with others, while Crawford et al. (2011) undertook a study examining the views of service users on traditional outcome measures. Such measures open the door to researchers using outcome measures that are highly valued by service users (Kabir and Wykes 2010).

Practical Barriers to Involving Service Users and Carers in Research

Along with the political barriers, there are several practical barriers, which may stand in the way of successful involvement. An obvious, but very significant point is that involvement can require considerable resources, in the form of either financial support or staff time. Although, outside of mental health, policies are very mixed, within the UK model of mental health involvement, it is widely accepted that involvees should be reimbursed for any expenses incurred. Having governmental support for involvement has proved to be very helpful in establishing this practice. Another resource that may well need to be supplied is the provision of training, so that involvees without a prior research background are able to understand the research process and make a constructive contribution.

Involvement might also require certain shifts in institutional practices, given the emphasis placed within universities on an established academic record as a prerequisite for a formal role in research activities. Both to ensure recruitment and to maximize diversity, researchers need to ensure that research opportunities for service users are widely advertised within appropriate contexts where they will come to the attention of service users, while both researcher and service user involvee need to maintain positive attitudes toward involvement if the enterprise is to have the best possible chance of success.[13] Finally, there is, of course, a strong possibility that involvees may become unwell at some point of the research process. There are a range of creative solutions for supporting users during periods of illness, which, understandably, tend to be worked out on a case-by-case basis, such as creating "action plans" in advance and involving multiple service users, so that input can continue if one involvee is unwell.[14]

CONCLUSION

Service user involvement in mental health provides a strong and growing body of evidence to support the incorporation of lived experience within medical research, while also challenging the models of knowledge and practice within conventional research on a number of different levels. Despite some challenges, it is a valuable innovation, which offers to improve the viability, relevance, and quality of research in various ways. At the same time, by elevating the mental health service user from the role of passive subject to active participant and "expert by experience," it provides an opportunity for the service user to perform a valued role within the research framework and community, and a growing amount of evidence shows the benefits of involvement to the involvee themselves, from altruism to gaining skills, experience, employment, a greater understanding of their illness, and so on.

The UK is very much leading the way on involvement in mental health research, with support at a governmental level making it an increasingly routine and expected element of research. Although some resources have been produced on how to achieve successful involvement, much work remains to be done. For example, involvement in basic medical research is arguably not as well developed as within more social science–oriented research. Of course, service user involvement in research requires, as does all research, significant investment of resources such as money and staff time. Nevertheless, where a sufficient commitment to involvement has been made, it is the incorporation of the knowledge by personal experience of mental illness that has allowed this research to achieve fuller aims, methods, inclusivity, and outcomes that could ever have been possible without the introduction of this distinct perspective.

NOTES

1. Throughout this chapter we refer to people with mental health problems as "service users," although many other terms are commonly used, including "patients," "clients," "consumers," and so on.

2. In relation to other conditions, lived experience critiques tend to focus more on insufficient attention or prioritization of their subjective experience of illness within the medical process, rather than a disagreement over the diagnosis or treatment itself. See, for example, Carel (2008); Toombs (2004).

3. There are some interesting parallels, for example in the UK, with rarely deployed powers of detention under the Public Health (Control of Disease) Act 1984 and Public Health (Infectious Disease) Regulations 1988. See, for example, Martin (2006) and also section 5 of http://www.peterbates.org.uk/uploads/5/5/9/5/55959237/vilebodies16.pdf (last accessed July 4, 2016).

4. See, for example, the World Health Organization's *Mental Health Atlas 2011* at http://apps.who.int/iris/bitstream/10665/44697/1/9799241564359_eng.pdf (last accessed July 4, 2016), for figures to demonstrate "substantial gap between the burden caused by mental disorders and the resources available to prevent and treat them."

5. http://psychiatrized.org/LeonardRoyFrank/FromTheFilesOfLeonardRoy-Frank.htm (last accessed July 4, 2016).

6. These included Diana Rose, who later became the world's first professor of user-led research at Kings College London.

7. See http://www.nihr.ac.uk/funding/public-involvement-in-your-research.htm (last accessed July 4, 2016).

8. Within the UK the expectation that service users will be involved in research mirrors the Health and Social Care Act (2012), which requires user involvement in government funded health services. See https://www.england.nhs.uk/wp-content/uploads/2013/09/trans-part-hc-guid1.pdf (last accessed July 4, 2016).

9. See http://www.invo.org.uk/wp-content/uploads/2012/06/INVOLVEevidenceresource.pdf (last accessed July 4, 2016).

10. See, for example, http://www.invo.org.uk/posttyperesource/why-should-members-of-the-public-be-involved-in-research (last accessed July 4, 2016).

11. In England state-provided mental health care is predominately provided by a number of regional authorities called "Trusts." (Each Trust is part of the UK-wide National Health Service or NHS.)

12. On the importance of neutral language for mental health involvement in general, see, for example, Roberts (2010).

13. In the UK ongoing attention has been given to how to address these practical issues. For a full treatment of how these barriers can be addressed, see http://www.invo.org.uk/wp-content/uploads/2012/04/INVOLVEBriefingNotesApr2012.pdf (last accessed July 4, 2016) and http://www.rds-sw.nihr.ac.uk/documents/NIHR_MHRN_Involving_Mental_Health_Problems_Research2013.pdf (last accessed July 4, 2016).

14. For some other strategies that have been used by institutions, see https://www.crn.nihr.ac.uk/wp-content/uploads/mentalhealth/UserCarerResearcherGuidelines-May2014_FINAL.pdf (last accessed July 4, 2016).

REFERENCES

Ashcroft, Joanne, Til Wykes, Joseph Taylor, Adam Crowther, and George Szmukler. 2016. "Impact on the Individual: What Do Patients and Carers Gain, Lose and Expect from being Involved in Research?" *Journal of Mental Health* 25(1): 28–35.

Baca-Garcia, Enrique, Maria Perez-Rodriguez, Ignacio Basurte-Villamor, Del Moral, Antonio L. Fernandez, Miguel Jimenez-Arriero, De Rivera, Jose L. Gonzalez, Jeronimo Saiz-Ruiz, and Maria A. Oquendo. 2007. "Diagnostic Stability of Psychiatric Disorders in Clinical Practice." *The British Journal of Psychiatry* 190(3): 210–216.

Bolton, Derek. 2009. "What is Mental Disorder?" *Psychiatry; Social, Ethical and Legal Aspects of Psychiatry* 8(12): 468–470.

Bonnie, R. J. 2002. "Political Abuse of Psychiatry in the Soviet Union and in China: Complexities and Controversies." *The Journal of the American Academy of Psychiatry and the Law* 30(1): 136–144.

Callard, Felicity, Diana Rose, and Til Wykes. 2012. "Close to the Bench as Well as at the Bedside: Involving Service Users in all Phases of Translational Research." *Health Expectations: An International Journal of Public Participation in Health Care and Health Policy* 15(4): 389–400.

Carel, Havi. 2008. *Illness*. Durham, U.K.: Routledge.

Charlton, James I. 1998. *Nothing about Us without Us: Disability Oppression and Empowerment*. Berkeley, California; London: University of California Press.

Clark, Carrie C., Elizabeth A. Scott, Katherine M. Boydell, and Paula Goering. 1999. "Effects of Client Interviewers on Client-Reported Satisfaction with Mental Health Services." *Psychiatric Services* 50(7): 961–963.

Corstens, Dick, Eleanor Longden, Simon McCarthy-Jones, Rachel Waddingham, and Neil Thomas. 2014. "Emerging Perspectives from the Hearing Voices Movement: Implications for Research and Practice." *Schizophrenia Bulletin* 40(Suppl 4): S285–S294.

Crawford, Mike J., Dan Robotham, Lavanya Thana, Sue Patterson, Tim Weaver, Rosemary Barber, Til Wykes, and Diana Rose. 2011. "Selecting Outcome Measures in Mental Health: The Views of Service Users." *Journal of Mental Health* 20(4): 336–346.

Cree, Lindsey, Helen L. Brooks, Kathryn Berzins, Claire Fraser, Karina Lovell, and Penny Bee. 2015. "Carers' Experiences of Involvement in Care Planning: A Qualitative Exploration of the Facilitators and Barriers to Engagement with Mental Health Services." *BMC Psychiatry* 15: 208.

Csipke, Emese, Constantina Papoulias, Silia Vitoratou, Paul Williams, Diana Rose, and Til Wykes. 2016. "Design in Mind: Eliciting Service User and Frontline Staff Perspectives on Psychiatric Ward Design through Participatory Methods." *Journal of Mental Health* (25)2: 4–21.

Dawson, John, and George Szmukler. 2006. "Fusion of Mental Health and Incapacity Legislation." *The British Journal of Psychiatry* 188(6): 504–509.

Ennis, Liam, and Til Wykes. 2013. "Impact of Patient Involvement in Mental Health Research: Longitudinal Study." *British Journal of Psychiatry* 203(5): 381–386.

Evans, Jo, Diana Rose, Clare Flach, Emese Csipke, Helen Glossop, Paul McCrone, Tom Craig, and Til Wykes. 2012. "VOICE: A New Measure of Service Users' Perceptions of Inpatient Care, using a Participatory Methodology." *Journal of Mental Health* 21(1): 57–71.

Fernyhough, Charles. 2014. "Hearing the Voice." *The Lancet* 384 (9948): 1090–1091.

Gergel, Tania Louise. 2014. "Too Similar, Too Different: The Paradoxical Dualism of Psychiatric Stigma." *The Psychiatrist* 38(4): 148–151.

Gillard, Steven, Rohan Borschmann, Kati Turner, Norman Goodrich-Purnell, Kathleen Lovell, and Mary Chambers. 2010. "What Difference Does it Make? Finding Evidence of the Impact of Mental Health Service User Researchers on Research into the Experiences of Detained Psychiatric Patients." *Health Expectations* 13(2): 185–194.

INVOLVE. 2012. *Briefing Notes for Researchers: Involving the Public in NHS, Public Health and Social Care Research*. INVOLVE, Eastleigh. http://www.invo.org.uk/wp-content/uploads/2014/11/9938_INVOLVE_Briefing_Notes_WEB.pdf (last accessed 4th July 2016).

Jones, Steven, Lee D. Mulligan, Heather Law, Graham Dunn, Mary Welford, Gina Smith, and Anthony P. Morrison. 2012. "A Randomised Controlled Trial of Recovery Focused CBT for Individuals with Early Bipolar Disorder." *BMC Psychiatry* 12: 204.

Kabir, Thomas, and Til Wykes. 2010. "Measures of Outcomes that are Valued by Service Users." In *Mental Health Outcome Measures*, edited by G. Thornicroft and M. Tansella. 3rd ed., London: Royal College of Psychiatrists Publications.

Lazarus, Lisa, Ashley Shaw, Sean LeBlanc, Alana Martin, Zack Marshall, Kristen Weersink, Dolly Lin, Kira Mandryk, Mark Tyndall, Mark, W., and the PROUD Community Advisory Group. 2014. "Establishing a Community-Based Participatory Research Partnership among People Who Use Drugs in Ottawa: The PROUD Cohort Study." *Harm Reduction Journal* 11: 26.

Martin, Robyn. 2006. "The Exercise of Public Health Powers in Cases of Infectious Disease: Human Rights Implications." *Medical Law Review* 14(1): 132–143.

Patterson, S., J. Trite, and T. Weaver. 2014. "Activity and Views of Service Users Involved in Mental Health Research: UK Survey." *The British Journal of Psychiatry* 205(1): 68–75.

Roberts, M. 2010. "Service User Involvement and the Restrictive Sense of Psychiatric Categories: The Challenge Facing Mental Health Nurses." *Journal of Psychiatric and Mental Health Nursing* 17(4): 289–294.

Rutter, Deborah, Catherine Manley, Tim Weaver, Mike J. Crawford, and Naomi Fulop. 2004. "Patients Or Partners? Case Studies of User Involvement in the Planning and Delivery of Adult Mental Health Services in London." *Social Science and Medicine* 58(10): 1973–1984.

Sanati, Abdi, and Michalis Kyratsous. 2015. "Epistemic Injustice in Assessment of Delusions." *Journal of Evaluation in Clinical Practice* 21(3): 479–485.

Semrau, Maya, Heidi Lempp, Roxanne Keynejad, Sara Evans-Lacko, James Mugisha, Shoba Raja, Jagannath Lamichhane, Atalay Alem, Graham Thornicroft, and Charlotte Hanlon. 2016. "Service User and Caregiver Involvement in Mental Health System Strengthening in Low- and Middle-Income Countries: Systematic Review." *BMC Health Services Research* 16(1): 79.

Simpson, Emma L., and Allan O. House. 2002. "Involving Users in the Delivery and Evaluation of Mental Health Services: Systematic Review." *BMJ* 325: 1265.

Svensson, Bengt, and Lars Hansson. 2016. "Satisfaction with Mental Health Services. A User Participation Approach." *Nordic Journal of Psychiatry* 60(5): 365–371.

Szmukler, George, Kristina Staley, and Thomas Kabir. 2011. "Service User Involvement in Research." *Asia-Pacific Psychiatry* 3(4): 180–186.

Toombs, S. Kay 2004. "Living and Dying with Dignity: Reflections on Lived Experience." *Journal of Palliative Care* 20(3): 193–200.

Trivedi, Premila, and Til Wykes. 2002. "From Passive Subjects to Equal Partners: Qualitative Review of User Involvement in Research." *The British Journal of Psychiatry* 181(6): 468–472.

Turner, Kati, and Steve Gillard. 2011. "'Still Out there?' Is the Service User Voice Becoming Lost as User Involvement Moves into the Mental Health Research Mainstream?" In *Critical Perspectives on User Involvement*, edited by Marian Barnes and Phil Cotterell, 189–199: Cambridge: Policy Press.

Chapter 6

The Role of Patient Perspectives in Clinical Case Reporting

Rachel A. Ankeny

Clinical cases have remained an exceedingly popular genre of publication in medicine due to their abilities to permit rapid and easy exchange of information particularly about unusual or difficult diagnostic, therapeutic, or prognostic puzzles. Despite this prominence, their status as sources of evidence or as the potential basis for generalizations remains contested. Although cases usually are presented with reference to a particular individual patient (or perhaps a small series of patients), the information provided typically is presented wholly from the point of view of the clinician and other healthcare practitioners with whom the patient has had contact. However, recent consensus guidelines formulated to standardize clinical case reporting encourage inclusion of the "patient perspective," that is, description of the case from the patient's point of view (Gagnier et al. 2013).

This chapter analyzes the use of patient perspectives as a source of evidence in published medical case reporting, with a particular focus on what ethical and epistemological values are supported by efforts to include this form of evidence, considered against the backdrop of general disquiet in some quarters about the status of cases particularly given the rise of evidence-based medicine (EBM). Inclusion of patient perspectives may help to fulfill various ethical responsibilities on the part of researcher-practitioners who publish cases; these duties can be argued to be due to patients who form the basis of a case, which is potentially a much more intense and personal type of publication (even when suitably anonymized) than publications based on groups of patients or similar. In epistemological terms, I argue that given that one key purpose of case reporting is to propose hypotheses and provide material for future research, rather than to establish definitive causal theories or similar, inclusion of patient perspectives allows incorporation of different types of evidence that might otherwise be overlooked in conventional case reporting.

Use of standpoint epistemology reveals that inclusion of patient perspectives has the potential to shift understandings of disease away from narrow biological processes to more holistic, socioculturally grounded understandings of wellness and deviations from it. They also permit a more epistemologically robust view of cases as epistemic objects rather than equivalent to their static, published instantiations. Thus, I propose that patient perspectives should be more widely encouraged and taken seriously as a key source of evidence within case reporting in medicine.

BACKGROUND: THE CASE IN MEDICINE

> Always note and record the unusual .... Publish it. Place it on permanent record as a short, concise note. Such communications are always of value. (Attributed to William Osler: see Thayer 1920, 51–2)

Cases are a critical form of medical publication: a survey in the early 2000s documented publication of approximately 40,000 new case reports per year, representing 13.5% of all publications in core medical journals (Rosselli and Otero 2002), and more recent analyses have claimed that the annual number of published cases continues to grow (Gagnier et al. 2013). Several peer-reviewed, open-access journals have been founded that are dedicated to cases (e.g., Kidd and Hubbard 2007) as have dedicated Internet-based databases for gathering and curating cases. Thus, the publication of cases is an increasingly popular form of communication of knowledge within medicine, despite the fact that the quality of case reporting has been noted to be highly uneven or even misleading (Richason et al. 2009; Kaszkin-Bettag and Hildebrandt 2012).

A published case report typically describes a medical problem experienced by a patient (or sometimes multiple patients), usually involving the presentation of an illness or a complaint that proved to be difficult for the clinicians involved to explain or categorize based on existing understandings of disease or understandings of physiology and pathology. Cases tend to take a highly standardized format (which is sometimes explicitly prescribed by the journal), outlining the presentation of the disease or complaint, diagnosis, treatment, and outcomes. Systematic guidelines for case reports recently have been developed using a consensus-driven international framework that aims to promote more consistency, transparency, completeness, and impact (Gagnier et al. 2013), and which will be discussed in more detail below.

Consider, for example, the following case (summarized from Plotnikoff 2014): a 56-year-old white, married, nondrinking, nonsmoking female health professional was referred for evaluation of 7 years of idiopathic recurrent acute pancreatitis requiring numerous hospitalizations. The patient's condition had been extensively explored using a range of investigative techniques and laboratory tests without any abnormal results except possible mild diffuse fatty infiltration of the liver. Recurrent pancreatitis can be life threatening; although it has a well-established differential diagnosis, no explanation for its occurrence currently exists and existing therapeutic options are highly invasive and often ineffectual. The patient was found to have marked elevation of fecal deoxycholic acid (DCA), a secondary bile acid used to induce pancreatitis in animal models. Once her fecal DCA levels normalized following treatment with antibiotics along with probiotics in order to balance the microbiome, she ceased having symptoms and hospitalizations, and had been symptom free for 48 months at the time of the publication of the case. The author argues that this case indicates that an observational study of fecal DCA levels is required, along with more attention to the microbiome in patients with recurrent acute pancreatitis.

As can be seen above, a main goal of case publication is to capture information including many details that may not be immediately relevant, but that could prove to be (e.g., the extensive demographics typically included). This approach permits the information contained in the case to be useful for other practitioners and researchers when a similar condition arises, and so that the case can be systematically combined with other cases into larger datasets. It also then can be used as the basis for more systematic examination of particular therapies or approaches, as is explicitly noted in the case summarized above. Case reports also can provide material and a framework for education in various healthcare settings, particularly given the rise of pedagogies based on problem- and case-based learning (pedagogical uses of cases are not discussed in this chapter; for a more extensive discussion of the various roles of cases, see Ankeny 2017).

Many cases depict complaints arising in specialty or subspecialty settings, particularly where diagnosis is difficult or tricky in some way, and describe uncommon or even "unique" clinical occurrences (McCarthy and Reilly 2000) or rare conditions (Carey 2010), observed under uncontrolled conditions (Simpson and Griggs 1985). Some report interventions proved useful, particularly in cases where randomized controlled trials (RCTs) of therapies would not be feasible due to low patient numbers and/or ethical issues (Albrecht et al. 2005), which can be particularly important for making comparisons across various settings.

Cases have sometimes been viewed pejoratively as too speculative and lacking in evidence, particularly because they often provide information about singular instances without any controls or underlying experimental methods, hence raising questions about their validity and generalizability. This type of criticism has become heightened due to the recent dominance of

EBM, which places the RCTs at the top of its hierarchy of evidence. Cases are typically observational, with the observations and their documentation being uncontrolled, as they occur within real-life clinical settings. However, those more sympathetic to the power of cases have noted that increasingly sophisticated methods for data analysis using natural language processing and techniques associated with "big data" should allow more rigorous evaluation of the data contained in published cases, "uncovering evidence hidden in what used to be regarded as anecdotes" (Riley 2013, 11).

Cases often are used to provide evidence for future investigations, given the rich detail typically contained in them. Hence, they frequently are reused as evidence of something other than that which they were originally intended to show, or serve as the basis for additional analysis with attention to different details than provided in the original publication in which they were presented, particularly because of the rich information that they typically contain (for more discussion on these processes and specific examples, see Ankeny 2011, 2014).

INCLUSION OF PATIENT PERSPECTIVES: PRINCIPLES AND MOTIVATIONS

Is there a person in this case? (Monroe, Holleman, and Holleman 1992)

Calls to include patient perspectives in various types of case presentations have occurred in medicine for at least the last 30 years. Critics have noted that the taking of a medical history in the context of patient care, which is typically the precursor to a published case, neglects the patient's ongoing life story and hence the meaning of the illness for the patient, thus serving as a dehumanizing force in medicine (Churchill and Churchill 1982). Various ongoing efforts particularly in medical pedagogy to include patient perspectives have resulted in clinicians writing in literary genres in order to speak from a patient-centered point of view using narrative (two classic examples of such efforts include Brody 1987 and Kleinman 1988); many medical journals include sections of such writings.

Such efforts, together with use of literature in the medical curriculum, have been noted to be useful for countering the hierarchical norms in medicine and the tendency to view patients as objects, in order to "empower patients to make decisions based on their values" and "level the playing field between physician and patient" (e.g., Calvert et al. 2013b). They also have been noted to be a means of instilling a deeper sense of the sociocultural effects of illness and disease in physicians, in part to make them better listeners and explainers (Charon 1986; Crawford 1992).

With regard to the construction of case histories in the context of an individual patient case, William J. Donnelly (1988; see also his 1986) urges that residents include the patient's understanding of his or her illness and its effects on the patient's life. The recent rise of patient "illness narratives" (see Frank 1995) has allowed patients to begin to write themselves back into the story of medicine (Wiltshire 1998). There are the occasional unusual examples of the patient point of view entering into published cases, though these are clearly exceptions and have been noted to be worthy of analysis due to their atypicality (see Hurwitz 2006).

More recently, the patient perspective has been introduced as an explicit component of consensus guidelines to standardize presentation of published cases. The CARE (CAse REport) Guidelines (Gagnier et al. 2013) were formulated to provide a framework to support consistency, transparency, and accuracy in the reporting of information from patient encounters and in the publication of case reports in particular. A consensus group utilized a standardized process for developing guidelines (see Moher et al. 2010), which included conducting a needs assessment, literature reviews, and interviews with experts to draft a checklist for case reporting, and the drafting of consensus-based case report guidelines that were then piloted, finalized, and published in 2013.

The CARE checklist includes a detailed summary of the minimal requirements for any published case and explicitly incorporates consideration of the patient's viewpoint, including his or her main concerns for seeking clinical care and, when appropriate, the patient's own perspective on or experience of his or her care presented in a brief narrative published with (or accompanying) the case report. The inclusion of patient perspectives is noted as having been discussed at length by the consensus group, as was the adoption of guidelines for such reporting, the latter of which ultimately were not formulated (Gagnier et al. 2013, 2). The group also noted the recent promulgation of an extension to the standardized methods for reporting patient-reported outcomes in RCTs, known as the CONSORT (Consolidated Standards of Reporting Trials) Statement, which are noted to be increasingly used to inform patient-focused care and clinical decision making, as well as health policy or reimbursement decisions (Calvert et al. 2013a).

Maintenance of confidentiality and informed consent generally are essential for publication of any case, as noted by the consensus group (Gagnier et al. 2013). Those preparing a case for publication are obligated to protect the anonymity of the patient(s) described in the case by removing details from the history or demographics that are not considered to be relevant and may allow identification by others. Statistics on how often patients' identities are compromised in medical literature are somewhat dated (Nylenna and Riis 1991), but underscore the need for caution. However, clinicians and others preparing material for publication, including cases, also must be careful not to alter essential information or falsify the reporting in the process of anonymization. For inclusion of the patient viewpoint in particular, it is considered to consult the patient or his or her guardian (or where impossible to do this, to seek permission from an institutional ethical committee or its equivalent) and to obtain explicit (rather than blanket) informed consent with reference to the actual content of the case publication itself prior to its publication.

TELLING CONFLICTING TALES IN MEDICINE VIA CASES

Doctors simplify; patients complicate. (Charon 1992, 116)

Many commentators have written on the complex nature of relationships between patients and doctors and how they use these to achieve common ground as part of the clinical encounter. Some contend that the relationship is best viewed as a partnership or form of collaboration, and thus that a critical component of this relationship is a commitment to shared meaning-making (e.g., Quill 1983; Suchman and Matthews 1988). Such shared meaning-making is the basis of effecting improvement of the patient's condition not only in the immediate term but on a longer basis because of the patient's active participation in the medical encounter.

Rita Charon (1992) has noted that analysis of the discourse associated with medical narratives reveals that patients and doctors in fact "are engaged in deep conflict about meaning and purpose" (116). Patients and doctors typically have deeply differing perceptions of the same events, with the medical encounter serving as a space within which they attempt to reconcile these perceptions. As Kathryn Montgomery Hunter (1991) notes, the stories told by each party are often conflicting and complicated by different agendas. Perhaps more importantly, they have fundamentally different goals and distinct views on the medical enterprise: "Their tasks are not parallel or complementary" (Charon 1992, 116).

Charon's emphasis on the different structures of discourse utilized by patients and clinicians is particularly relevant for the purposes of considering the integration of patient perspectives into published cases. As she notes, physicians use words "to contain, to control, and to enclose" (ibid.), whereas patients tend to be more inclusive and much less selective. What perhaps is particularly notable in the instance of a medical case is that the doctor's voice becomes authoritative, even though in a sense his or her version of the events could be seen as a mere interpretation of the "real" case as narrated by the patient. Charon describes this as the "gloss" on the part of the doctor as compared to the "text" itself produced by the patient, and notes that this results in an inversion of typical power relations. However, undoubtedly there is a privileging of the doctor's voice over that of the patient, particularly in the case of compiling a case for the purposes of publication.

Another aspect of published cases is that a certain version of an ideal is assumed by the narrator-clinician, a norm of health or well-being from which the patient has deviated; Charon (1992) associates this ideal with classical and traditional notions of the perfect form (119). In philosophical terms, what is critical here is that an essential part of constructing a published case is determining what norm applies, since any account of causality is closely tied to certain conceptions of deviations from this norm, which may differ depending on the underlying theory of disease adopted. In contrast, the patient's priority is not understanding this norm, or the underlying theory of disease associated with it, but having his or her health restored, that is, wanting to feel better.

Finally, the clinician's goals in writing up a particular patient encounter for publication are at least twofold: one is to recount his or her actions and to validate them through imposition of the usual medical rules that are well recognized by the peers likely to read the account (cf. the role of mortality and morbidity meetings as recounted by Bosk 1979). But publication also allows the clinician-author to bring attention to this particular case as an example of something that might be seen again, in other words to make a specific instance into a generic one, or at least one that might be applicable to someone else, somewhere else, at a different point in time. In contrast, the patient does not view himself or herself as an instance of something broader, but an individual with a narrative rich in specific and personal details of life. For his or her purposes, it matters little if this illness or complaint represents something generalizable (it may well be the case that clinician-authors are in fact conflicted about this aspect of cases too, but this is not my focus in this chapter: for a linguistic-based analysis of this issue, see Salager-Meyer, Alcaraz Ariza, and Briceño 2010).

WHAT DO WE GAIN? ETHICAL AND EPISTEMIC ADVANTAGES

Cases are found, cases are objects, cases are made, and cases are conventions. (Ragin and Becker 1992, 9–10)

Although advocates of inclusion of the patient point of view in medical encounters have emphasized a number of critical points about why such perspectives are important to include, as outlined above, this section makes explicit several additional advantages that have not been addressed in any detail in the available literature, with particular focus on published cases. First, inclusion of patient perspectives may help to fulfil various ethical responsibilities on the part of researcher-practitioners who publish cases. Certain duties can be argued to be due to patients who form the basis of a case, which is potentially a much more intense and personal type of publication (even when suitably anonymized) than publications based on groups of patients or similar. The patient who becomes the basis of a published case often has been extensively scrutinized due to having a puzzling, dramatic, and/or confusing medical presentation, and typically has been subjected to a larger number of tests and other forms of monitoring than more routine patients due to the difficulties faced in proposing a diagnosis or finding an effective therapy. Rather than further reinforcing the power differential inherent in this type of situation, particularly given the historic tendency not to seek permission from those who became the basis of published cases, encouraging inclusion of the patient perspective, in close consultation with the patient himself or herself (or those who represent the patient, in the case of children or other dependents) and with his or her consent, can allow the patient's voice to be heard not only for the purposes of the current clinical encounter but for those who subsequently consult and utilize the case.

Including patient perspectives also assists in reorienting the focus about what cases are: by allowing the patient to become a joint author of the text by contributing his or her viewpoint and voice, published cases can be more firmly grounded in the complexities of the illness or clinical problem at hand. This emphasis can focus the reader on what the case is a case *of*, rather than solely on a particular patient (see discussion of a similar point in the context of clinical case histories by Monroe, Holleman, and Holleman 1992, following on Charon 1989). This sort of "thick description" (to use Clifford Geertz's 1973 term) also potentially provides much more material for the published case with which others can work, including data that might not initially seem relevant but that could prove to be important over time.

An example of this can be seen in the published case reporting associated with the published case description of what came to be known as toxic shock syndrome (analyzed in more detail in Ankeny 2011). Although difficult now for us to imagine, there was no mention in the original published case (Todd et al. 1978; see also Morris 1989) of how many of the affected patients were female and what their menstrual status was, as the focus instead was on the novel, acute, and severe disease processes caused in these female (and male) pediatric patients who had been affected by a toxin-producing strain of *Staphylococcus*, which seemed similar to historic instances of childhood scarlet fever. Being more cognizant of basic demographic and health information from the patient's point of view (in this case including tampon usage, the correlation of which with the occurrence of the illness was noted in subsequent studies using retrospective case analysis; see Davis et al. 1980; Shands et al. 1980) might well have allowed earlier detection of some of the key underlying causal factors in this case.

Given that one key purpose of case reporting is to propose hypotheses and provide material for future research (as occurred, for instance, in the example of the case of the woman with pancreatitis provided at the start of this chapter), inclusion of patient perspectives allows incorporation of different types of evidence that might otherwise be overlooked in conventional case reporting. Patients tell their stories in rich detail, hoping to assist the doctor to figure out what is wrong and how to fix it. Among the information typically included by patients are their values, experiences, and theories and guesses about causality, including how their own behaviors or other stimuli may have contributed to the symptoms or disease in question. There also are rules of causality underlying published cases, rules that are known and assumed by the clinician-author and his or her likely readers who are other clinicians and researchers typically seeking information or guidance about a similar case. However, these understandings of causality, their likelihood, and even claims about what is biologically possible are often not shared by the patient and may well be in tension with those assumed by healthcare practitioners.

Thus, in the typical published case that does not include patient perspectives, a particular point of view is assumed or marked off especially in terms of possible causal structures, and others are necessarily disregarded. In contrast, inclusion of patient perspectives may well allow other views to be incorporated that are unlikely but could be relevant not only for articulation of causes but for some other aspects of the case in question when reanalyzed by future reader-users. This type of argument is supported more generally by calls for pluralistic models of clinical reasoning (e.g., Tonelli 2006) that incorporate multiple types of warranted evidence into decision making, including less-traditional sources of evidence such as patient perspectives that operate alongside more traditional or conventional sources of data (such as RCTs).

At the most fundamental epistemic level, including patient perspectives in published cases helps to resituate cases in the broader context within which they originally were observed and documented. To modify an argument made with reference to data sharing in medical research and elsewhere (Mauthner and Perry 2013), cases are inseparable from the context within which they are generated. Cases in their real clinical form are not bounded or black-boxed entities, although they become such (perhaps of necessity) when written and published, and hence made static. To use Karin Knorr-Cetina's terms (2001), cases as epistemic objects produced through scientific inquiry have the capacity to unfold indefinitely .... They are more like open drawers filled with folders extending indefinitely into the depth of a dark closet. Since epistemic objects are always in the process of being materially defined, they continually acquire new properties and change the ones they have. (181)

Given that a main purpose of publishing a case is to allow others to build on the material contained in it, often drawing on that which is provided but often is not the main focus (see Ankeny 2011, 2014), the cumulative knowledge generated from a case similarly relies on an iterative process that causes new properties to emerge and even new theories and explanations to emerge based on what was provided in the original published case. Patient perspectives can be one critical part of this process, particularly because they often bring in additional information that may not immediately be relevant but that could prove useful over time given that such information is less bounded by rules or particular understandings of disease.

Use of standpoint epistemology reveals that inclusion of patient perspectives has the potential to shift understandings of disease away from narrow biological processes to more holistic, socioculturally grounded understandings of wellness and deviations from it. Standpoint theories emphasize that our understandings of the world typically take a privileged point of view, and hence we should be more explicit about the specific, socially situated perspectives from which claims to epistemic privilege or authority are made (e.g., Haraway 1988). Such considerations are particularly important in relation to topics or issues that are socially or politically contested, and for groups that have been systematically disadvantaged: in this instance, the perspectives of those who are unwell could be considered to be necessary to integrate into consideration in order to counter tendencies to privilege other voices, notably those of healthcare professionals. In turn, patient perspectives may allow access to issues that might otherwise be neglected, for instance, relating to the sociocultural and other nonbiological drivers of disease and health, thus displacing the interpretive consensus that otherwise is typically imposed by the voiceless and highly standardized narration of more traditional published cases.

As Brian Hurwitz (2006) has written with reference to an unusual case that includes considerable detail about the patient's broader social concerns that help to frame her illness, these perspectives can create "degrees of ambiguity and possibility" (235). If we accept that clinical reasoning involves weighing up a range of types of evidence and creating an "interpretative story" (Tanenbaum 1994), patient perspectives are a critical part of this narrative creation and telling when it is done via published cases. In turn, integration of patient perspectives thus can contribute to ongoing critiques of EBM, particularly given its well-recognized positivist underpinnings and its privileging of certain forms of evidence (Harari 2001; Goldenberg 2006, 2015).

Those features of illness and disease that are often excluded as subjective and nonrigorous are precisely those that can be reinstilled in published cases via patient perspectives and which in fact are arguably essential to the treatment of disease in many instances (Toombs 1993).

As an example of several of these potential contributions of patient perspectives to published cases, consider a recently published case (Kienle et al. 2013) written using the CARE guidelines (and included on their website as a model for case publication): a 10-year-old girl with type 1 diabetes mellitus along with a developmental disorder with motor and sensory dysfunction required full-time care, which was provided by her mother and an assistant nurse. Despite insulin therapy, she had poor glucose control with frequent mild and severe hypoglycemic episodes. She also was socially isolated, not able to participate in activities with her peers, and generally frustrated with her situation. In turn, the case describes the mother as substantially stressed and the parent-child relationship as endangered. The team referred the patient and her mother to a multidisciplinary diabetic care program with emphasis on individualized and patient-centered care, including focus on self-management. Through a shift away from technical management and laboratory data to the young girl and her specific problems, wishes, goals, needs, and conditions from her point of view, the child is described as having become an active partner in management and decisions and also able to successfully self-manage her disease, resulting in substantial improvement not only in glucose control but also in her quality of life and relationship with her mother. This example not only represents inclusion of a patient perspective (at least implicitly) as part of the published case, but also emphasizes a change of focus away from narrowly biomedical outcomes to the broader sociocultural and psychological context within which the disease is occurring, and hence the team was able to foster a more effective treatment that simultaneously aligned with the patient's own goals and values.

CONCLUSION

Clearly, the inclusion of patient perspectives is no simple task, as evidenced by the fact that relatively few cases published to date since the issuance of the CARE Guidelines in 2013 have explicitly contained discussion of them as a recommended option. In part this lacunae may simply be due to a type of learning curve but also may reflect the relative lack of emphasis on the patient viewpoint in the guidelines as well as the absence of explicit, detailed guidelines about how to elicit such a viewpoint and include it within a published case, what the key components of the patient viewpoint might be considered to be, how to seek informed consent, and so on, in marked contrast to the details provided on the obligatory components of any published case.

On a more philosophical level, a case that includes a patient perspective will necessarily be a more complex entity than the usual published case written solely from the clinician's point of view. As Kathryn Montgomery Hunter (1988) has noted, all stories must have a teller, who is the sole provider of information and thus, regardless of his or her reliability, retains a certain sense of authority. The hybrid case that contains both the usual clinician-author narrative but also the patient viewpoint may well be thought of as too complicated for future potential readers and lacking in guidance about how to understand the patient's point of view (often still written in the clinician's own voice) and what to do with the information provided by the patient.

Perhaps one way to reconcile this tension is to view the inclusion of patient perspectives as underscoring the idea that published cases are a form of history. As such, despite attempts at standardization via the CARE Guidelines and other efforts, they are neither completely coherent in the manner of a (well-crafted) fictional narrative nor as epistemologically secure as proofs or experimental replications made using the scientific method (Montgomery 1992). Hence, the patient viewpoint is yet one more aspect of what already tends to be a messy, rather than neat, package of information (see Ankeny 2011), gathered in part because a particular patient's clinical presentation is confusing and puzzling to the practitioners confronted by it.

A final point about why inclusion of patient perspectives in published cases is important is that it may contribute to growing tendency not to merely privilege certain forms of research in medicine (those prioritized by EBM, for instance) but to develop deeper understandings of appropriate, effective, and rigorous ways of using other methodologies such as observation: "When it becomes clear how new data contributes to evidence, the stewardship needed to produce high-quality data will become more rewarding and our attitude toward 'observation' will shift .... This will transform how we think about 'evidence' and revolutionize its creation, diffusion, and use—opening new opportunity landscapes" (Gagnier et al. 2013, 2). Hence, as this chapter has claimed, patient perspectives should be given a more central role in published cases (and detailed guidelines for their inclusion should be developed) for a range of ethical and epistemological reasons: we can recognize the problems and potential limitations of this type of evidence without in any way denying that it nonetheless may have epistemic value and that a pluralistic model of clinical reasoning is warranted.

ACKNOWLEDGMENTS

Many thanks to Michael Kidd of Flinders University and Editor of *Journal of Medical Case Reports* for discussions about this journal, the CARE Guidelines, patient perspectives, and case reporting in general. I also gratefully acknowledge participants at the European Research Council grant-funded workshop "What Is Data-Intensive Science?" held in December 2014 at the University of Exeter (see www.datastudies.eu), particularly Sabina Leonelli and Mary Morgan, for comments on a related paper that influenced the arguments in this chapter.

REFERENCES

Albrecht, Jörg, Alexander Meves, and Michael Bigby. 2005. "Case Reports and Case Series from *Lancet* had Significant Impact on Medical Literature." *Journal of Clinical Epidemiology* 58: 1227–32.

Ankeny, Rachel A. 2011. "Using Cases to Establish Novel Diagnoses: Creating Generic Facts by Making Particular Facts Travel Together." In *How Well Do Facts Travel? The Dissemination of Reliable Knowledge,* edited by Peter Howlett and Mary S. Morgan (eds.), 252–72. Cambridge: Cambridge University Press.

Ankeny, Rachel A. 2014. "The Overlooked Role of Cases in Casual Attribution in Medicine." *Philosophy of Science* 81: 999–1011.

Ankeny, Rachel A. 2017 (forthcoming). "The Case in Medicine." In *The Routledge Companion to Philosophy of Medicine,* edited by Miriam Solomon, Jeremy R. Simon, and Harold Kincaid. New York: Routledge.

Bosk, Charles. 1979. *Forgive and Remember: Managing Medical Failure.* Chicago: University of Chicago Press.

Brody, Howard. 1987. *Stories of Sickness.* New Haven: Yale University Press.

Calvert, Melanie et al. for the CONSORT PRO Group. 2013a. "Reporting of Patient-Reported Outcomes in Randomized Trials: The CONSORT PRO Extension." *Journal of the American Medical Association* 309: 814–22.

Calvert, Melanie et al. 2013b. "The CONSORT Patient-Reported Outcome (PRO) Extension: Implications for Clinical Trials and Practice." *Health and Quality of Life Outcomes* 11: 184.

Carey, John C. 2010. "The Importance of Case Reports in Advancing Scientific Knowledge of Rare Diseases." In *Rare Diseases Epidemiology,* edited by Manuel Posada de la Paz and Stephen C. Groft, 77–86. Dordrecht: Springer.

Charon, Rita. 1986. "To Render the Life of Patients." *Literature and Medicine* 5: 58–74.

Charon, Rita. 1989. "Doctor-Patient/Reader-Writer: Learning to Find the Text." *Soundings* 72: 137–52.

Charon, Rita. 1992. "To Build a Case: Medical Histories as Traditions in Conflict." *Literature and Medicine* 11: 115–32.

Churchill, Larry R. and Sandra W. Churchill. 1982. "Storytelling in Medical Arenas: The Art of Self-Determination." *Literature and Medicine* 1: 77.

Crawford, T. Hugh. 1992. "The Politics of Narrative Form." *Literature and Medicine* 11: 147–62.

Davis, Jeffrey P. et al. 1980. "Toxic Shock Syndrome: Epidemiological Features, Recurrence, Risk Factors, and Prevention." *New England Journal of Medicine* 303: 1430.

Donnelly, William J. 1986. "Medical Language as Symptom: Doctor Talk in Teaching Hospitals." *Perspectives in Biology and Medicine* 30 (Autumn): 81–94.

Donnelly, William J. 1988. "Righting the Medical Record: Transforming Chronicle into Story." *Journal of the American Medical Association* 260: 825.

Frank, Arthur W. 1995. *The Wounded Storyteller: Body, Illness and Ethics*. Chicago: University of Chicago Press.

Gagnier, Joel J. et al. for the CARE Group. 2013. "The CARE Guidelines: Consensus-Based Clinical Case Reporting Guideline Development." *Journal of Medical Case Reports* 7: 223, DOI: 10.1186/1752-1947-7-223 (published in numerous journals simultaneously as an open-access publication).

Geertz, Clifford. 1973. *The Interpretation of Culture*. New York: Basic Books.

Goldenberg, Maya. 2006. "On Evidence and Evidence-Based Medicine: Lessons from the Philosophy of Science." *Social Science and Medicine* 62: 2621–32.

Goldenberg, Maya J. 2015. "Whose Social Values? Evaluating Canada's 'Death of Evidence' Controversy." *Canadian Journal of Philosophy* 45: 404–24.

Harari, Edwin. 2001. "Whose Evidence? Lessons from the Philosophy of Science and the Epistemology of Medicine." *Australian and New Zealand Journal of Psychiatry* 35: 724–30.

Haraway, Donna. 1988. "Situated Knowledges: The Science Question in Feminist and the Privilege of Partial Perspective." *Feminist Studies* 14: 575–99.

Hunter, Kathryn Montgomery. 1988. "Making a Case." *Literature and Medicine* 7: 74–5.

Hunter, Kathryn Montgomery. 1991. *Doctor's Stories: The Narrative Structure of Medical Knowledge*. Princeton: Princeton University Press.

Hunter, Kathryn Montgomery. 1992. "Remaking the Case." *Literature and Medicine* 11: 163–79.

Hurwitz, Brian. 2006. "Form and Representation in Clinical Case Reports." *Literature and Medicine* 25: 216–40.

Kaszkin-Bettag, Marietta and Wolfram Hildebrandt. 2012. "Case Reports on Cancer Therapies: The Urgent Need to Improve the Reporting Quality." *Global Advances in Health and Medicine* 1: 8–10.

Kidd, Michael and Charlotte Hubbard. 2007. "Introducing *Journal of Medical Case Reports*." *Journal of Medical Case Reports* 1: 1.

Kienle, Gunver S. et al. 2013. "Patient-Centered Diabetes Care in Children: An Integrated, Individualized, Systems-oriented, and Multidisciplinary Approach." *Global Advances in Health and Medicine* 2: 12–9.

Kleinman, Arthur. 1988. *The Illness Narratives: Suffering, Healing and the Human Condition*. New York: Basic Books.

Knorr-Cetina, Karin. 2001. "Objectual Practice." In *The Practice Turn in Contemporary Theory*, edited by Theodore R. Schatzki, Karin Knorr-Cetina, and Eike von Savigny, 175–88. London: Routledge.

Mauthner, Natasha Susan and Odette Parry. 2013. "Open Access Digital Data Sharing: Principles, Policies and Practices." *Social Epistemology* 27: 47–67.

McCarthy, Laine H., and Kathryn E.H. Reilly. 2000. "How to Write a Case Report." *Family Medicine* 32: 190–95.

Moher, David et al. 2010. "Guidance for Developers of Health Research Reporting Guidelines." *PLoS Medicine* 7(2): e1000217.

Monroe, William F., Warren L. Holleman, and Marsha C. Holleman. 1992. "Is There a Person in this Case?" *Literature and Medicine* 11: 45–63.

Morris, B. A. 1989. "The Importance of Case Reports (Letter to the Editor)." *Canadian Medical Association Journal* 141: 875–6.

Nylenna, Magne and Povl Riis. 1991. "Identification of Patients in Medical Publications: Need for Informed Consent." *British Medical Journal* 302: 1182.

Plotnikoff, Gregory A. 2014. "Elevated Deoxycholic Acid and Idiopathic Recurrent Acute Pancreatitis: A Case Report with 48 Months of Follow-up." *Global Advance in Health and Medicine* 3: 70–72.

Quill, Timothy. 1983. "Partnerships in Patient Care: A Contractual Approach." *Annals of Internal Medicine* 98: 228–34.

Ragin, Charles C. and Howard S. Becker. 1992. *What is a Case?* Cambridge: Cambridge University Press.

Richason, Tiffany P. et al. 2009. "Case Reports Describing Treatments in the Emergency Medicine Literature: Missing and Misleading Information." *BMC Emergency Medicine* 9: 10.

Riley, David. 2013. "Case Reports in the Era of Clinical Trials." *Global Advances in Health and Medicine* 2: 10–11.

Rosselli, Diego and Andres Otero. 2002. "The Case Report is Far from Dead (Letter)." *The Lancet* 359: 84.

Salager-Meyer, Françoise Elisabeth, María Ángeles Alcaraz Ariza, and Marianela Luzardo Briceño. 2010. "Titling and Authorship Practices in Medical Case Reports: A Diachronic Study." *Communication and Medicine* 10: 63–80.

Shands, Kathryn N. et al. 1980. "Toxic-Shock Syndrome in Menstruating Women: Association with Tampon Use and Staphylococcus aureus and Clinical Features in 52 Cases." *New England Journal of Medicine* 303: 1436–42.

Simpson, Ross J. and Thomas R. Griggs. 1985. "Case Reports and Medical Progress." *Perspectives in Biology and Medicine* 28: 402–6.

Suchman, Anthony and Dale Matthews. 1988. "What Makes the Doctor-Patient Relationship Therapeutic? Exploring the Connexional Dimension of Medical Care." *Annals of Internal Medicine* 108: 125–30.

Tanenbaum, Sandra. 1994. "Knowing and Acting in Medical Practice: The Epistemological Politics of Outcomes Research." *Journal of Health, Politics, Policy, and Law* 19: 27–44.

Thayer, W. S. 1920. *Osler, The Teacher Sir William Osler, Bart.* Baltimore: Johns Hopkins Press.

Todd, James et al. 1978. "Toxic-shock Syndrome Associated with Phage-Group-I-Staphylococci." *Lancet* 2: 1116–18.

Tonelli, Mark R. 2006. "Integrating Evidence into Clinical Practice: An Alternative to Evidence-based Approaches." *Journal of Evaluation in Clinical Practice* 12: 248–56.

Toombs, S. Kay. 1993. *The Meaning of Illness: A Phenomenological Account of the Different Perspectives of Physicians and Patients.* Dordrecht: Springer.

Wiltshire, John. 1998. "The Patient Writes Back: Bioethics and the Illness Narrative." In *Renegotiating Ethics in Literature, Philosophy, and Theory,* edited by Jane Adamson, Richard Freadman, and David Parker, 181–98. Cambridge: Cambridge University Press.

Chapter 7

"The Science is Clear!"

Media Uptake of Health Research into Vaccine Hesitancy

Maya J. Goldenberg and Christopher McCron

While the movement from research to practice in medicine and health policy is well studied in the philosophy of medicine, an underresearched component of this knowledge-to-action trajectory has been the influence of media reporting on newsworthy health research. Media analysis has characteristically been a focus of communications and not philosophical research. However, the epistemic and rhetorical impact of science and health reporting warrants attention by philosophers as another facet of the complex science-values relationship in healthcare. Science journalism [of which health makes up roughly 50% of its content (Hargreaves 2012)] shapes public understanding and engagement, which in turn influences patient choice as well as organizational and policy decisions. All the while, science reporting is frequently criticized for sensationalizing new findings and omitting the uncertainty of novel research, thereby misleading the publics[1] into thinking that new scientific ideas are fully established (see, for example, Abola and Prasad 2016). In this chapter, we highlight many of these issues in our presentation of an original case study of media uptake of scientific research addressing a politically charged issue: vaccine hesitancy.

In early 2014, news media jumped on the pre-print publication of a study into the effectiveness of vaccine information messaging in the journal *Pediatrics* by political scientists Brendan Nyhan and colleagues. The study measured the effect of a variety of pro-vaccine messages designed to reduce vaccine misperception and increase rates of childhood vaccination against measles, mumps, and rubella (MMR) by vaccine-hesitant parents. Media and public interest was strong as this timely study was published in the same year that the United States experienced a 20-year high in incidences of measles—a worrisome situation that was attributed to geographical pockets of willfully unvaccinated children (Pugh 2014). These outbreaks occurred despite several

113

decades of active public health efforts to counter vaccine fears among the general population. Many observers despaired over this potentially disastrous public health threat.

While it is typically positive research findings that are sensationalized, this study was picked up enthusiastically by the press for its *negative* findings: none of the tested vaccine messages were effective in improving attitudes toward immunization among the subsample of vaccine-hesitant parents. Indeed, a "backfire effect" was noted—some vaccine-hesitant parents responded to the corrective information by becoming *less* likely to vaccinate their children. A fatalistic conclusion was quickly drawn by the press; headlines read "You Can't Change an Anti-Vaxxers Mind" (Mooney 2014a) and "Nothing, Not Even Hard Facts, Can Make Anti-Vaxxers Change Their Minds" (Alter 2014).

This conclusion made for a gripping news story at a time when public frustration over vaccine-refusing parents was high. The news coverage suggested that anti-vaxxers were not only irresponsible and selfish (which was the popular thinking on the subject), but were now also incapable of reason. This media rush also came out at a time when many states and other jurisdictions were considering controversial measures to end personal belief exemptions for school-aged children. These allegedly conclusive scientific findings that people could not be convinced to vaccinate their children were offered in the press to bolster this and other liberty-limiting measures on parental choice.

All the while, this fatalistic conclusion did not follow from the research. Our analysis of the news coverage of Nyhan et al.'s research reveals frequent mischaracterizations of the research findings, which made for a more interesting news story that was used to promote political ends. This case study underscores the importance of careful media communications in order to promote public understanding of science by illustrating the trajectory of misinformation that can arise when communication channels are poorly maintained.

MEDIA SHAPING OF PUBLIC UNDERSTANDING OF SCIENCE

It is somewhat surprising to find that the influence of the media on the medical research-to-practice trajectory is an underresearched area of philosophy of medicine, given the field's attentiveness to the science-values interface. Health journalism is the access point for most of the publics to engage with health research and policy initiatives; the journalistic narratives shape understanding and attitudes, which influence patient choice as well as organizational and policy decisions.

Communications research offers some important insights. News media is understood to possess an "agenda-setting function" (McCombs and

Shaw 1972) insofar as journalism directs public attention to issues (see McCombs and Shaw 1972; Rogers, Dearing and Chang 1991) and frames those issues (Entman 2007; 2010). When news media assigns responsibility for causing and fixing social problems, it can inform the judgments and actions of citizens and policymakers (Iyengar 1996; Kim and Willis 2007; Weiner 2006).

Communications studies also offers a small body of research addressing how media shapes public understanding of science. The most comprehensive effort comes from Hargreaves et al.'s (2012) recent study, *Toward a Better Map: Science, the Public, and the Media*, which undertook media content analysis of climate change, MMR vaccine, and genetic research coverage in the British press in 2002. The team also conducted two nationally representative surveys in 2002 of over 1,000 adults in Britain to measure their understanding of those same issues relative to their news consumption habits. The researchers' aim was to find out "what people knew about those science issues regularly in the news, and where there is public interest involved" (Hargreaves et al. 2012).

Hargreaves et al. (2012) were able to conclude that news media plays a role in informing people's understanding of scientific issues. However, those inputs are incomplete. Survey respondents retained the main themes or frameworks of news coverage of science-related stories, while (sometimes important) informational details were not remembered. The research team proposed that those themes or frameworks are used as building blocks for people to make sense of an issue. While these frames sometimes allowed news consumers to make informed guesses about those issues, they could also generate misunderstandings. People will use whatever information they have to make sense of the world, often filling in "gaps" with unfounded beliefs. Related research corroborates the previous finding that most people consume news inattentively, thereby taking in only certain aspects of a story. The information that *does* stick is often based on repeated associations (Lewis 2001).

Thus, the media's framing of controversial issues like vaccine safety warrants scrutiny—what gets said, how it is said, and what gets left out. The framing of news stories has significant public impact in terms of general understanding and mobilizing action. Prior coverage of the MMR vaccine controversy demonstrates this well. Indeed, the media shouldered some blame for lowered vaccination rates in Europe due to the poor handling of the initial MMR-autism scare that exploded into the papers in 1998 with the publication of the infamous study by Dr. Andrew Wakefield et al. (1998) alleging a causal link (Holton et al. 2012; see also Begg et al. 1998). Science journalist Seth Mnookin's (2011) review of the media uptake surrounding the publication recounts the London press's enthusiastic uptake of the "lone wolf" narrative of a humble doctor fighting the establishment, a story that was animated by emotional accounts of mothers, struggling to raise children with autism, who were now convinced that the MMR vaccine had caused their children's conditions. The ample data already mounted against Wakefield's claims and credibility received far less mention in the papers.

The media also courted the vaccine-autism controversy by actively reporting both sides of the debate. While offering both sides of the story is a journalistic norm—a practice meant to encourage "fair and balanced" reporting—this standard of best journalistic practice came under fire for paradoxically misleading the public by giving the impression that the two sides held equal weight (Dixon and Clarke 2012; Clarke 2008; Lewis and Speers 2003; Offit and Coffin 2006). The two sides were misconstrued as there was insufficient mention of how marginal the vaccine skeptical position was against the weighty scientific consensus that vaccines had no correlation to autism. And where the strong consensus on vaccine safety was mentioned, it did not receive a suitable level of prominence insofar as it was typically offered as *post hoc* reassurance to the newsworthy concerns of parents (Lewis and Speers 2003). To illustrate:

> The following [news reporting] examples are typical: "The government has mounted campaigns to persuade parents the MMR jab is safe after some research linked it to autism and bowel disorders in children" and "Ministers continue to insist the MMR jab, which some doctors have linked to autism, is the best way of protecting children." (ibid.)

A content analysis of 279 news articles in U.S. and UK papers from 1998 to 1999 found the majority to offer so-called balanced reporting of the MMR-autism link. Specifically, reporters presented the opposing claims that supported and refuted a connection and often left the reader to decipher who to believe (Clarke 2008). Cunningham (2003) has noted that reporters are often discouraged from adjudicating competing claims or providing sufficient context for readers to assess the claims. To editorialize can be perceived as bias (worthy of opinion pieces rather than news reporting) and a failure on the part of the reporter to be "objective" in her journalism. That standard of "objectivity" has become heavily contested and is commonly debated in journalism ethics forums (Cunningham 2003; Part IV of Meyers 2010; chapter 4 of Ward 2011).

With characteristically inattentive news consumers repeatedly hearing associations of vaccines and autism in the framing of vaccine safety as controversial, it should not be surprising that Hargreaves et al. (2012) found many of their survey respondents to believe that vaccines cause autism. They also found that efforts made by the press to deny that association, that is, by reporting that the link has not been established, had only a minimal effect.

Broadsheet readers, who had more exposure to criticism of the vaccine-autism association than tabloid readers, were only slightly more likely to deny the claim that vaccines cause autism. In the end, "what people appear to have heard was simply that there were two sides to the debate" (Hargreaves et al. 2012). Thus, it was the broad pattern of media coverage, specifically the framing of the MMR-autism controversy as a legitimate debate, rather than the informational details, that were effectively communicated. This led 54% of Hargreaves et al.'s (2012) October 2002 survey respondents to think that "since both sides got roughly equal coverage, they must correspond to roughly equal bodies of evidence" (ibid.).

And so, media communications have created bias by way of balanced reporting.[2] Additionally, the press must heavily consider the collateral of reporting "public interest" stories. Health reporters often encounter health scares and exciting "breakthroughs." Hargreaves et al. (2012) advised cautious communications in light of their investigation into MMR-autism coverage:

In brief, while Wakefield's claims are of legitimate public interest, our report does give credence to the view that research questioning the safety of something that is widely used should be approached with caution, both by scientists publishing that research and journalists covering it. This is especially the case if any decline in public confidence has negative consequences for public health.

Health reporters have reflected on the fall-out from news coverage of the Wakefield controversy, alleging that they have learned important lessons about careful reporting. But with the uncertainty of emerging science and the pressure of tight newsroom deadlines, that cautious approach proves difficult to exercise. Reporter Julia Belluz (2015b) has reflected on how the Wakefield incident "scared the media away from covering vaccine side effects."[3] She cites the underreporting of rare cases of narcolepsy following Pandemrix influenza vaccinations in Europe in 2009–2010 (see Centers for Disease Control and Prevention 2015) as illustrative of this problem for health reporters. Instead, she writes, there is pressure for reporters to be vaccine "cheerleaders" (Belluz 2015b). This challenge not only underscores how difficult it can be to report on controversial research topics but also highlights how important it is to strike that difficult balance of responsible reporting. While journalists are held to standards of accuracy, there is no meaningful way to report "just the facts" (indeed, there is no such thing), and it is not clear how much responsibility journalists must hold for passive news consumption habits that make readers prone to misunderstanding important issues. These considerations deserve more time and attention than we can offer here.

THE CASE: "EFFECTIVE MESSAGES IN VACCINE PROMOTION: A RANDOMIZED TRIAL"

"Effective Messages in Vaccine Promotion: A Randomized Trial" (Nyhan et al. 2014) tested the effectiveness of several typical vaccine-promoting messages used by public health agencies to persuade parents. The researchers concluded that no single vaccine message effectively motivated parents to vaccinate their children. In fact, there was a measurable "backfire effect" insofar as some parents became *less* inclined to have their children immunized after exposure to the promotional materials (Nyhan et al. 2014).

Two online surveys were administered using a nationally representative sample of parents over the age of 18 with one or more children younger than 17 ($n = 1759$). While the first survey established baseline measures, such as current attitudes toward vaccines and health, the second survey randomly assigned parents into one of four different interventions or a control group. In each of these interventions, parents were exposed to a specific type of message that is commonly used in vaccine health communication, while the control group was exposed to a message about the pros and cons of bird feeding. The first intervention, called "autism correction," aimed to debunk the myth that vaccines cause autism by presenting scientific evidence that refuted the vaccine-autism link. A second intervention, entitled "disease risks," presented information about the symptoms and risks associated with diseases prevented by the MMR vaccine. "Disease narrative," the third intervention, told a dramatic story of a child nearly dying from measles. The fourth and final intervention was "disease images," which featured photos of young children suffering from measles, mumps, or rubella. To increase the generalizability of the results, the autism correction, disease risks, and the disease narrative were all adapted from the Center for Disease Control and Prevention's (CDC) website nearly word-for-word. However, the source information of the CDC materials was withheld from the parents in order to avoid having them tie their interpretations of the materials to prior views about the organization. In wave one of the study, subjects read the intervention materials. In wave two, Nyhan and colleagues measured three dependent variables of interest: (1) parents' perceptions of whether the MMR vaccine could cause autism in a healthy child, (2) parents' opinions of the likelihood of a child suffering serious side effects from an MMR vaccine, and (3) parents' intentions to have their own child vaccinated.

The final results of the study disappointingly showed that none of the interventions increased parent's intentions to have their child vaccinated (see Table 7.1). Also, when parents were informed about the risks of preventable diseases, it did not significantly affect their perceptions regarding the apparent risks of autism or serious vaccine side effects. The autism correction

Table 7.1 Results from Nyhan et al. 2014

Intervention	Belief that Vaccines Cause Autism	Fear of Vaccine Side Effects	Intention to Vaccinate
Autism correction	Lowered	Same	Same and Lowered ("backfire effect")
Disease risks	Same	Same	Same
Disease narrative	Same	Increased	Same
Disease images	Increased	Same	Same

intervention succeeded in reducing the perceived risks of a healthy child becoming autistic, but it did not create a significant reduction in concern for other serious side effects. Furthermore, some of the findings showed that the interventions could be detrimental in some ways. For instance, the disease narrative intervention increased parent's concerns regarding the risk of serious side effects from vaccines. Paradoxically, among the parents that were initially most opposed to vaccination, the autism correction decreased belief in autism being caused by vaccines, but it also further reduced their intentions to vaccinate their own child. While 70% of the control group's most hesitant parents claimed to be "very likely" to vaccinate their children, only 45% of the equally hesitant parents who had been exposed to the autism correction claimed to be "very likely" to vaccinate. Nyhan and colleagues referred to this phenomenon as a "backfire effect"; when confronting evidence that debunked the vaccine-autism link, "respondents brought to mind other concerns about vaccines to defend their anti-vaccine attitudes" (Nyhan et al. 2014, p. 840). The researchers figured that parents employed motivated reasoning,[4] a cognitive process driven by the desire to avoid cognitive dissonance, whereby subjects might "move the goalpost" (Haelle 2014) or develop other elaborate rationalizations in order to justify maintaining their prior beliefs about vaccines being unsafe.

Media and public interest in the research were strong amidst widespread public alarm over vaccine refusal and disease outbreaks. Because public health outreach had been ineffective in its concerted efforts to sway public opinion on vaccine safety (Goldenberg 2016; Macdonald et al. 2012; Black and Rappouli 2010), there was strong interest in radically changing the parameters of the discourse. A review of the media uptake suggests that this research was regarded as a means for doing so.

AN ANALYSIS OF MEDIA REPORTS ON THE STUDY

In order to study media uptake of Nyhan et al.'s (2014) MMR vaccine communications study, we searched Google News for news items that referenced either the study or Nyhan. We collected until saturation[5] was achieved. Our examination of 36 news items reporting directly on Nyhan et al.'s (2014) MMR messaging study or citing it in the context of other vaccine-related story lines revealed a consistent media focus on two aspects of the research findings: (1) that none of the vaccine interventions achieved the intended goal of increasing parents' intention to immunize their children and (2) that such efforts could backfire insofar as some vaccine hesitators became less likely to comply with vaccine recommendations. The reports were negative in tone (see Abrams 2014; Aleccia 2014; Mooney 2014a; Konnikova 2014; Bouie 2015; Stafford 2015; Selbig 2015) and sometimes fatalistic in the claim that the study demonstrated that *nothing* would change an anti-vaccinator's mind (Abrams 2014; Aleccia 2014; Alter 2014; Mooney 2014a). These anxiety-inducing story lines, with provocative headlines like "Trying to Convince Parents to Vaccinate Their Kids Just Makes the Problem Worse" (Aleccia 2014), misrepresented the study's findings. The conclusion that nothing could change a vaccine-hesitant parent's mind did not follow from the study results, nor was it suggested in the discussion section of the publication.[6] Only four specific messages were tested, three of which came from the same anonymized source (the CDC). And while they did represent typical messaging employed by public health agencies, the researchers made no suggestion that their sample represented the gamut of vaccine communications strategies.

A number of other articles stopped short of such a fatalistic conclusion that *nothing* worked but still pushed the claims of the study too far. For example, one news headline stated "Study Shows Michigan's Vaccine Education Program Could Backfire" despite obvious differences between the Michigan program and the interventions tested by Nyhan et al. (2014). The Michigan program involved conversation with a public health nurse for parents requesting philosophical exemptions for vaccines (Selbig 2015). The CDC interventions that had been tested were, in contrast, textual in format and were not delivered by a trusted medical professional.[7] Furthermore, the study had not evaluated parental attitudes toward philosophical exemptions specifically. The reporter proposed that it *may* even be impossible to change the minds of "anti-vaxxers," a claim that, while only suggestive, is not supported by the study's findings. Other news reports overextended the study findings to claim that people with anti-vaccine views were immune to *facts*, *science*, and *reason*. King (2015) misinterpreted the study findings by suggesting that the backfire effect was experienced by all parents in the study. He also misunderstood a statement made by Nyhan in an interview— "Throwing facts and evidence at people rarely changes their minds, particularly when it comes to issues we care about"—to mean that vaccine hesitators cannot be educated. This led him to ask rhetorically, "But if facts can't fight anti-vaccination myths, then what can?" Similarly, *The Conversation* ran the headline "Throwing science at anti-vaxxers just makes them more hardline" (Stafford 2015). Bouie (2015), writing in *Slate*, added that "reason doesn't work either." The study had not, of course, examined facts, science, or reason.

These sorts of extreme claims would often be preceded by factual errors in the reporting of the study; these errors lent credibility to those conclusions. There were several instances of conflation of textual information with other forms of persuasive communication such as conversation (Selbig 2015; Mooney 2014a; Wolfe 2015). The study did not test verbal persuasion, yet Wolfe (2015) reports that Nyhan et al.'s findings show the backfire effect could occur in the context of conversation with one's doctor (Wolfe 2015). Nyhan and colleagues actually suggest that pediatricians might be our best hope for persuading vaccine-hesitant parents (Nyhan et al. 2014; Nyhan in Barton 2014; Nyhan in Tremonti 2015; Nyhan in Belluz 2015a).

Another common error was misrepresenting the sample population, or failing to report that only the participants with the most negative vaccine attitudes experienced the backfire effect (see Bernstein 2015; King 2015; Johnson 2015; Wolfe 2015). This omission creates the false impression that the entire study sample experienced some degree of the backfire effect. Another news item incorrectly reported that *all* interventions caused a backfire effect (French 2015), when in fact hardened anti-vaccine views arose only among some of the participants in three intervention arms. Other journalists oversimplified the study findings with such statements as "They found that, when you tried to use evidence to make a case, it backfired: Anti-vaccination convictions deepened" (Johnson 2015). Instead, different interventions had different effects. The combination of negativity and fatalism, reporting errors, and extreme wording tended to detract from and distort Nyhan and colleague's final conclusion that "these results suggest the need to carefully test vaccination messaging before making it public" (Nyhan et al. 2014, p. 841). Instead, the resonant message captured in the media coverage was that the hope of effective pro-vaccine communications had been proven (nearly or completely) futile.

One year later, we saw similarly overblown press coverage of a new study by Nyhan and Reifler (2015) testing flu shot communications that yielded similar findings to the MMR vaccine study. Once again, media coverage conveyed dismay over dire findings. This study investigated the myth that the flu shot can give you the flu and how it influenced participants' intentions to vaccinate. Again, the study surveyed a nationally representative sample in two waves before and after reading CDC materials that debunked the flu shot myth. Among participants with the highest concerns of serious vaccine side effects, the flu myth intervention reduced misconceptions surrounding the influenza vaccine, but it also decreased the participants' intentions to vaccinate against the flu in the future. Nyhan and Reifler had succeeded in replicating the backfire effect.

In media coverage of the flu study, journalists adopted similar extreme wording and somber conclusions (for example, Mooney 2015b). For instance, one reporter asked, "If correcting misinformation doesn't work—what does?" (Romm 2015). Another wrote, "It's nigh impossible to change hearts and minds on vaccines" (Ingraham 2015). Similar to the MMR study, the flu shot study had tested only a few CDC textual interventions and had not concluded that all corrective information did not work. Yet readers were informed that "it seems that telling people the truth about common misconceptions about the flu vaccine is actually a bad idea" (Burks 2015). News reports were similarly prone to factual errors in their coverage of the study and offered misleading headlines. For instance, "Disproving Flu Vaccine Myths Doesn't Convince People to Get Vaccinated" (Burks 2014) and "Debunking Vaccine Junk Science Won't Change People's Minds" (Belluz 2015a). These titles read as if the findings of the flu shot study had definitively disproven the efficacy of corrective information. Textual information used in the study was also similarly conflated with face-to-face conversation (Burks 2014). One author declared that "a recent study found that, when vaccine-fearing patients heard the real facts *from their doctors*, they were actually more reluctant to get their flu shots" (Strauss 2014; emphasis added). There were, of course, no doctors involved in the flu shot study. In a similar manner, Ingraham (2015) claimed that the correction to the flu myth resulted in "reinforced views," which is the opposite of what the study found. Participants' misconceptions of the flu shot were in fact corrected; their intention to vaccinate, however, decreased.

In closing, the media coverage of Nyhan and colleague's vaccine communications studies was marked by frequent reporting errors and misleading extreme wording and the fatalistic conclusion that nothing can be done to convince vaccine skeptics (see Table 7.2). In all of the news articles reviewed, only one discussed the limitations of either MMR or flu study (Oxenham 2015), an important means for having the reader understand the full implications of the study (see Table 7.2). These narratives made for a provocative story arc in the context of high public anxiety over vaccine-preventable disease outbreaks occurring in the United States at the time. Concurrent to the public fears amidst worrisome measles outbreaks in 2014 and 2015 (Centers for Disease Control and Prevention 2016) was heated public debate over proposed policy measures—some radical—to curtail vaccine resistance. We will now further argue that some of the reports of the study findings were enlisted to spuriously defend policy positions on vaccine exemptions. The research findings were thereby manipulated to serve political ends.

Table 7.2   Summary of Findings from Media Analysis of 36 News Items Reporting on Nyhan et al. (2014) Study

Negative or fatalistic tone	17/36[a]
Misrepresented study findings (i.e., contained reporting errors, misrepresented backfire effect)	15/36[b]
Presented limitations of the study	1/36[c]
Discussed legislative action	10/36[d]
Uncategorized	7/36[e]

[a]Abrams 2014; Aleccia 2014; Alter 2014; Bouie 2015; Burks 2014; Haelle 2014; Ingraham 2015; Konnikova 2014; Mooney 2014a; Mooney 2014b; Page 2015; Romm 2014; Sanders 2015; Selbig 2015; Singal 2015; Stafford 2015; Strauss 2014.
[b]Abrams 2014; Aleccia 2014; Bernstein 2015; Bouie 2015; Belluz 2015a; Burks 2014; French 2015; Ingraham 2015; Johnson 2015; King 2015; Mooney 2014a; Selbig 2015; Stafford 2015; Strauss 2014; Wolfe 2015.
[c]Oxenham 2015.
[d]Alferis 2015; Bouie 2015; Forman 2015; Haelle 2015; Jackson 2015; Johnson 2015; Peeples 2015; Sanders 2015; Selbig 2015; Wolfe 2015.
[e]Barton 2014; Diamond 2015; Graves 2015; Joss 2015; Maron 2015; McKay and Whalen 2015; Tremonti 2015.

ADVANCING POLITICAL ARGUMENTS

Within this context of increased vaccine-preventable disease outbreaks (Centers for Disease Control and Prevention 2016) and heightened public anger over the actions of the vaccine-refusing minority (Healy and Paulson 2015; Aldhous 2015a; Aldhous 2015b), 2015 became an active year for legislators to challenge their state's nonmedical exemption laws permitting personal belief/philosophical and religious grounds for vaccine refusal.[8] Legislators came to see that where nonmedical exemptions had previously assisted in maintaining strong vaccine programs by placating vaccine resistance, these exemptions, particularly the slightly ambiguous "personal belief" allowance, had become a problem. In California, for instance, personal belief exemptions had doubled since 2007, thereby creating the conditions for numerous whooping cough and measles outbreaks within the state in recent years (Majumder et al. 2015; Mello et al. 2015). These legislative efforts to curtail nonmedical exemptions would not be easy, but the political climate was right for dramatic measures to delimit individual parental choice in support of enforced measures to protect the common good.

While the fight in California was the most high profile—due in part to the national attention received by the "Disneyland" outbreak that affected people in 17 states and also Senate Bill 277's dramatic reform measures—no less than 18 states introduced various bills to either tighten requirements, make the process more cumbersome, publicize school immunization rates, or, most drastically, eliminate some forms of vaccine exemptions (National Conference of State Legislatures 2016). President Obama, Speaker of the House John Boehner, and Republican leadership candidates Chris Christie and Paul Rand publicly commented on these legislative efforts, thereby bringing the issue of vaccine legislation into sharper public focus. Media pundits joined in the heated debates over personal freedoms versus community protection, with some drawing on Nyhan et al.'s research to bolster their arguments.

While appealing to relevant scientific research in the face of political fracas—not to resolve the debate but to assist in evaluating some of the competing claims (Goldenberg 2016)—is a credible journalistic practice, misappropriating those findings to bolster pet theories falls outside of the ethical norms of accuracy and accountability for journalists (see, for example, Canadian Association of Journalists 2015; American Society for Newsroom Editors 1975). Nyhan et al.'s research was at times mischaracterized or misapplied in order to bolster journalists' own positions regarding vaccine legislation.

In our data set, 10 out of 36 articles included some discussion of the warrants and efficacy of stricter rules surrounding personal exemption. Bouie's (2015) mischaracterization of Nyhan et al.'s research finding that reasoning with vaccine hesitator does not work was used to support the view that coercion was thereby justified. He writes that he would like to persuade vaccine-hesitant parents on the safety and efficacy of vaccines, but points to the research in anticipation of failed effort. He concludes, "If persuasion doesn't work, then I'm OK with coercion, too." Had Bouie consulted Nyhan on this conclusion or paid attention to public statements made by the researcher on this topic (Nyhan in Aliferis 2015; Nyhan in Forman 2015; Nyhan 2015), he would have found Nyhan to be unsupportive of this legislative effort. In those statements, Nyhan refers to some policy experts' warnings that mandatory vaccination laws are likely to backfire by further galvanizing anti-vaccine sentiment and that less coercive measures may be more effective (Ropiek 2015; Omer et al. 2012). Bouie did not consult any of that research into the merits of coercive public health measures. Coercion was a foregone conclusion given his mischaracterization of vaccine-hesitant parents.

Forman (2015) and Aliferis (2015), in contrast, utilized Nyhan et al.'s research to support an alternative legislative measure to mandatory vaccination. Forman drew on the team's positive finding that informing parents that vaccines do not cause autism *did* successfully reduce belief in the myth itself as well as the finding that parents with the least favorable attitudes became *less* likely to vaccinate their children, to support the "nudge" approach of increasing the administrative burden placed on parents seeking nonmedical exemptions. The thinking behind nudge theory (Thaler and Sunstein 2008) is that nonincentivizing an unwanted behavior will create nonforced compliance more effectively than direct instruction or enforcement.[9] The more effort that needs to be put into acquiring a vaccine exemption, the less likely it is for parents to exercise that choice. Research confirms this: studies comparing vaccine legislation and rates of exemption show that American states with more cumbersome administrative requirements for ascertaining nonmedical exemptions have far lower use of this exemption (Omer et al. 2012).[10] Aliferis (2015) represents the research accurately, using it, along with other sources in favor of the "nudge" approach, to present a one-sided endorsement of administrative procedures over mandatory vaccination requirement.

Nyhan has expressed support for nudge efforts like requiring face-to-face education interventions with one's physician in order to obtain a signed exemption (in Forman 2015). He explains that even if the education does not convince the listener, the additional burden of having to visit a doctor could be motivating enough to change behavior. Among the mildly hesitant or time-constrained population, the difference between, say, clicking a box on a screen and meeting with a health care provider to obtain a signed exemption document can change vaccine compliance. With heavier administrative burdens in place, vaccination rates can rise without stoking public anger by eliminating exemptions entirely.

Reviewing the remaining articles utilizing Nyhan's research to promote policy positions, we found more misappropriation of the research findings. Selbig (2015), discussed earlier, offered a misreading of Nyhan's research in order to argue against Michigan's educational effort to increase vaccine compliance. Johnson (2015) overstates Nyhan et al.'s research conclusions regarding the ineffectiveness of persuasion techniques in order to support legislative efforts to change behavior.

Thus, the frequent mischaracterizations of Nyhan et al.'s research findings seemed to do much more than offer an exciting narrative for a competitive news market; they were utilized to promote political ends. Amidst heated debate over restricting nonmedical exemptions for school- and daycare-entry vaccines, and thereby curtailing parents' right to choose, the fatalistic reporting of the research being demonstrative that *nothing* could change an anti-vaccinator's mind was utilized to support controversial legislation.

CONCLUSION

As philosophy of medicine and science increasingly explores the values at play not just within scientific knowledge formation, but also in the interpretation and uptake of research findings in the clinical, organizational, and policy context, more attention is being given to the complex social terrain in which science and the publics meet. It is with this in mind that we presented an original case study into media uptake of recent vaccine communications research where we tracked the construction of the dismal conclusion that "You Can't Change an Anti-Vaxxer's Mind."

Our review of media uptake of Nyhan et al.'s 2014 MMR vaccine communications study found negative framing of the findings, with emphasis on what certain communications *failed* to achieve with respect to behavioral changes, rather than what had succeeded (i.e., diminished misperception of an vaccine-autism link), and how future communications could be improved in light of these findings. Some news reports drew the inaccurate headline-grabbing conclusion that no intervention could change beliefs and behaviors. This was the message that would most likely be remembered by the readership, as news consumers are known to retain only the amplified features of news stories.

This shocking finding that *nothing* could convince vaccine hesitators otherwise made for an exciting story line at a time when public frustration over vaccine refusal was high. Further, these news reports surfaced as political momentum was growing in favor of controversial legislation to restrict or eliminate nonmedical exemptions for school-entry vaccine requirements. The alleged conclusiveness of the scientific findings lent support to these efforts by discrediting persuasion techniques as a viable alternative to enforcement and arguably made the harms associated with restricting parental rights more palatable by framing those "anti-vaxxer" parents as incapable of reasonable judgment. The case for the common good trumping individual liberties was easier to make in the face of an irrational group of people.

Media institutions play significant roles in mediating the science-publics interface. We have focused on the roles played by news organizations in engaging and educating the publics. Other important research has attended to the equally weighty task of keeping scientific and governmental institutions accountable (for example, Mulgan 2003). When provocative headlines like "The Science is Clear: Anti-Vaxxers Are Immune to the Truth" (Editors 2015) appear in a national broadsheet amidst a climate of successive disease outbreaks and increasing public anger directed at "anti-vaxxers," the media is serving to inflame rather than inform the publics. Doing this keeps the publics from being able to engage meaningfully in informed discourse regarding issues that are important to us all.

NOTES

1. Science and communications studies prefer use of the term "the publics" instead of "the public" or "public sphere" in order to deny the notion that there is a unified body of lay people that interacts singularly with expert science. Instead, there are a plurality of nonexpert modes of engaging with science.

2. The same criticism of "balance as bias" has been made in the context of news coverage of the so-called climate debate in the United States (see Boycoff and Boycoff [2004], Oreskes [2004], and Malka et al. [2009]).

3. See also Borel (2015) on being criticized for raising concern about the financial conflicts of interests held by Kevin Folta, a prominent scientist and vocal GMO advocate.

4. See Mooney (2011).

5. Saturation is a term used in qualitative research to refer to the point where nothing new appears through further collection of data.

6. Nor did Nyhan, the primary researcher, ever raise that possibility that no vaccine communication intervention would work in the considerable media he did in surrounding the publication.

7. Research has found medical professionals to be the most trusted source of vaccine information for parents (Omer et al. 2009; Freed et al. 2011). In interviews, Nyhan, the primary investigator, recommended conversations with physicians as a good vaccine intervention on numerous occasions in interviews regarding the study (Tremonti 2015; Barton 2014; Belluz 2015a).

8. Despite being available in almost all U.S. states, national records show religious exemptions make up less than half of nonmedical exemptions. This is because they are, in many states, more difficult to obtain than other nonmedical exemptions. Parents may be required to cite and explain the religious doctrine in question. States with philosophical exemptions have 2.5 times the rate of exemptions than states with only religious exemptions (Vestal 2015). The more widely utilized personal belief/philosophical exemption was only available in 20 states in 2015. Starting June 2016, only 18 states will permit personal belief exemptions, with legislation successfully passed in California and Vermont to strike that option. California defined "personal belief exemption" broadly to include religious exemptions, thereby making the state's bill the most sweeping from all other reform measures introduced in 2015.

9. To define the term "nudge," Thaler and Sunstein write: "A nudge, as we will use the term, is any aspect of the choice architecture that alters people's behavior in a predictable way without forbidding any options or significantly changing their economic incentives. To count as a mere nudge, the intervention must be easy and cheap to avoid. Nudges are not mandates. Putting fruit at eye level counts as a nudge. Banning junk food does not" (Thaler and Sunstein 2008).

10. In what appears to be unfortunate oversight, it was easier for Californians to obtain a vaccine exemption than to obtain a child's vaccine records for school entry until some changes to legislation were enacted in 2014 (Nyhan 2015).

REFERENCES

Aliferis, Lisa. 2015. "Beyond abolishing the 'Personal belief exemption' to raise vaccination rates." *KQED News*, February 24. Accessed May 11, 2015. http://ww2.kqed.org/stateofhealth/2015/02/24/beyond-abolishing-the-personal-belief-exemption-to-raise-vaccination-rates/

Aldhous, Peter. 2015a. "These professors want parents who don't vaccinate to pay a tax." *Buzzfeed*, February 9. Accessed February 3, 2016. http://www.buzzfeed.com/peteraldhous/these-professors-want-parents-who-dont-vaccinate-to-pay-a-ta#.hdG01jpd2

Aldhous, Peter. 2015b. "This "pro-vaccine" doc has enraged the medical mainstream." *Buzzfeed*, February 13. Accessed February 3, 2016. http://www.buzzfeed.com/peteraldhous/this-pro-vaccine-doc-has-enraged-the-medical-mainstream#.lmBQ5BdNx

Alter, Charlotte. 2014. "Nothing, not even hard facts, can make anti-vaxxers change their minds." *Time*, March 4. Accessed May 7, 2015. http://healthland.time.com/2014/03/04/nothing-not-even-hard-facts-can-make-anti-vaxxers-change-their-minds/

Abrams, Lindsay. 2014. "Study: Trying to convince parents to vaccinate their kids just makes the problem worse." *Salon*, March 3. Accessed May 6, 2015. http://www.salon.com/2014/03/03/study_trying_to_convince_parents_to_vaccinate_their_kids_just_makes_the_problem_worse/

Aleccia, Jonel. 2014. "Pro-vaccine messages actually backfire, study finds." *NBC News*, March 3. Accessed May 7, 2015. http://www.nbcnews.com/health/health-news/pro-vaccine-messages-actually-backfire-study-finds-n41611

American Society of News Editors. 2016. "Statement of principles." Accessed February 3, 2016. http://asne.org/content.asp?pl=24&sl=171&contentid=171

Barton, Adriana. 2014. "Safety of vaccines is still a tough sell for some parents, study finds." *The Globe and Mail*, March 3. Accessed May 7, 2015. http://www.theglobeandmail.com/life/health-and-fitness/health/safety-of-vaccines-is-still-a-tough-sell-with-some-parents-study-finds/article17163515/

Begg, Norman, Mary Ramsay, Joanne White, and Zoltan Bozoky. 1998. "Media dents confidence in MMR vaccine." *British Medical Journal* 316 (7130): 561.

Belluz, Julia. 2015a. "Debunking vaccine junk science won't change peoples' minds. Here's what will." *Vox*, February 7. Accessed May 7, 2015. http://www.vox.com/2015/2/7/7993289/vaccine-beliefs

Belluz, Julia. 2015b. "How anti-vaxxers have scared the media away from covering vaccine side effects." *Vox*, July 27. Accessed January 20, 2015. http://www.vox.com/2015/7/27/9047819/H1N1-pandemic-narcolepsy-Pandemrix

Bernstein, Sharon. 2015. "Softer, less strident outreach may help calm U.S. vaccine skeptics." *Reuters*, February 11. Accessed May 11, 2015. http://www.reuters.com/article/us-usa-measles-vaccinations-idUSKBN0LF15E20150211

Borel, Brooke. 2015. "The problem with science journalism: We've forgotten that reality matters most." *The Guardian*, December 30. Accessed February 3, 2016. http://www.theguardian.com/media/2015/dec/30/problem-with-science-journalism-2015-reality-kevin-folta

Bouie, Jamelle. 2015. "How to deal with anti-vaxxers." *Slate*, February 2. Accessed May 12, 2015. http://www.slate.com/articles/news_and_politics/politics/2015/02/anti_vaxxers_resist_persuasion_if_they_refuse_we_have_to_force_them_to_vaccinate.html

Burks, Robin. 2014. "Disproving flu vaccine myths doesn't convince people to get vaccinated." *Tech Times*, December 12. Accessed May 7, 2015. http://www.techtimes.com/articles/21987/20141212/disproving-flu-vaccine-myths-doesnt-convince-people-to-get-vaccinated.htm

Canadian Association of Journalists. 2011. "Ethical guidelines." Accessed February 3, 2016. http://www.caj.ca/ethics-guidelines/

Centers for Disease Control and Prevention. 2015. "Narcolepsy following pandemrix influenza vaccination in Europe." Accessed February 3, 2016. http://www.cdc.gov/vaccinesafety/concerns/history/narcolepsy-flu.html

Centers for Disease Control and Prevention. 2016. "Measles cases and outbreaks." Accessed February 3, 2016. http://www.cdc.gov/measles/cases-outbreaks.html

Clarke, Christopher. 2008. "A question of balance: The autism-vaccine controversy in the British and American elite press." *Science Communication* 30 (1): 77–107.

Cunningham, Brent. 2003. "Rethinking objectivity." *Columbia Journalism Review* July/Aug 2003. http://www.cjr.org/feature/rethinking_objectivity.php

Diamond, Dan. 2015. "Ok, Measles is a problem. Which one of these 7 strategies will solve it?" *Forbes*, February 3. Accessed May 11, 2015. http://www.forbes.com/sites/dandiamond/2015/02/03/vaccine-deniers-seven-rational-proposals-to-solve-americas-irrational-problem/#1b592d0bf812

Dixon, Graham, and Christopher Clarke. 2012. "The effect of falsely balanced reporting of the autism–vaccine controversy on vaccine safety perceptions and behavioral intentions." *Health Education Research* 28 (2): 352–59.

Editors. 2015. "The science is clear: anti-vaxxers are immune to the truth." *The Globe and Mail*, February 13. Accessed February 3, 2016. http://www.theglobeandmail.com/opinion/editorials/the-science-is-clear-anti-vaxxers-are-immune-to-the-truth/article22987563/

Entman, Robert M. 2007. "Framing bias: Media in the distribution of power." *Journal of Communication* 57 (1): 163–73.

Entman, Robert M. 2010. "Media framing biases and political power: Explaining slant in news of Campaign 2008." *Journalism* 11 (4): 389–408.

Forman, Laura. 2015. "Experts question need for new legislation over belief-based immunization exemptions." *Peninsula Press*, April 6. Accessed May 11, 2015. http://peninsulapress.com/2015/04/06/legislation-belief-based-immunization-exemptions/

Freed, Gary L., Sarah J. Clark, Amy T. Butchart, Dianne C. Singer, and Matthew M. Davis. 2011. "Sources and perceived credibility of vaccine-safety information for parents." *Pediatrics*, 127 (1): 107–12.

French, Kristen C. 2015. "Fighting anti-vaxxers with a marathon science reading." *The Verge*, April 22. Accessed May 7, 2015. http://www.theverge.com/2015/4/22/8467559/fighting-anti-vaxxers-with-a-marathon-reading

Goldenberg, Maya J. 2016. "Public misunderstanding of science? Reframing the problem of vaccine hesitancy." *Perspectives on Science* 24 (5): 552–81.

Graves, Chris. 2015. "What we can learn from the vaccine wars." *PR Week*, March 2. Accessed May 11, 2015. http://www.prweek.com/article/1336135/learn-vaccine-wars

Haelle, Tara. 2014. "Debunking vaccine myths can have an unintended effect." *NPR*, December 11. Accessed May 6, 2015. http://www.npr.org/sections/health-shots/2014/12/11/369868202/debunking-vaccine-myths-can-have-a-surprising-effect

Haelle, Tara. 2015. "15 myths about anti-vaxxers, debunked – part 3." *Forbes,* February 19. Accessed May 12, 2015. http://www.forbes.com/sites/tarahaelle/2015/02/19/15-myths-about-anti-vaxxers-debunked-part-3/#31c1d62170be

Holton, Avery, Brooke Weberling, Christopher E. Clarke, and Michael J. Smith. 2012. "The blame frame: Media attribution of culpability about the mmr–autism vaccination scare." *Health Communication* 27 (7): 690–701.

Ingraham, Christopher. 2015. "Chart: New measles cases skyrocketed in January." *The Washington Post,* February 2. Accessed May 11, 2015. https://www.washingtonpost.com/news/wonk/wp/2015/02/02/there-were-more-new-measles-cases-in-the-past-month-than-in-all-of-2012/

Jackson, Harry R. Jr. 2015. "The disease debate of 2015." *Townhall.com,* February 17. Accessed May 12, 2015. http://townhall.com/columnists/harryrjacksonjr/2015/02/17/the-disease-debate-of-2015-n1958089/page/full

Johnson, Nathanael. 2015. "How to talk to an anti-vaxxer." *Grist,* February 4. Accessed May 11, 2015. http://grist.org/politics/how-to-talk-to-an-anti-vaxxer/

Joss, Laurel. 2015. "Shaming anti-vaccine parents could backfire, experts warn." *Autism Daily Newscast,* February 5. Accessed May 11, 2015. http://www.autismdailynewscast.com/shaming-anti-vaccine-parents-backfire-experts-warn/22736/laurel-joss/

King, Robin Levinson. 2015. "Inside the mind of anti-vaxxers." *The Star,* February 23. Accessed May 7, 2015. https://www.thestar.com/life/2015/02/23/inside-the-mind-of-anti-vaxxers.html

Konnikova, Maria. 2014. "I don't want to be right." *The New Yorker,* May 16. Accessed May 7, 2015. http://www.newyorker.com/science/maria-konnikova/i-dont-want-to-be-right

Lewis, Justin. 2001. *Constructing Public Opinion: How Political Elites Do What They Like and Why We Seem to Go Along With It.* New York: Columbia University Press.

Lewis, Justin, and Tammy Speers. 2003. "Misleading media reporting? The MMR story." *Nature Reviews: Immunology* 3 (11): 913–18.

Majumder, Maimuna S., Emily L. Cohn, Sumiko R. Mekaru, Jane E. Huston, and John S. Brownstein. 2015. "Substandard vaccination compliance and the 2015 measles outbreak." *JAMA Pediatrics* 169 (5): 494–95.

Maron, Dina Fine. 2015. "How to get more parents to vaccinate their kids." *Scientific American,* February 19. Accessed May 11, 2015. http://www.scientificamerican.com/article/how-to-get-more-parents-to-vaccinate-their-kids/

Meyers, Christopher. 2010. (ed.) *Journalism Ethics: A Philosophical Approach.* New York: Oxford University Press.

McCombs, Maxwell E., and Donald L. Shaw. 1972. "The agenda-setting function of mass media." *Public Opinion Quarterly* 36 (2): 176–87.

McKay, Betsy, and Jeanne Whalen. 2015. "Doctors work to ease vaccine fears." *The Wall Street Journal,* February 9. Accessed May 11, 2015. http://www.wsj.com/articles/doctors-work-to-ease-vaccine-fears-1423447481

Mello, Michelle M., David M. Studdert, and Wendy E. Parmet. 2015. "Shifting vaccination politics – the end of personal-belief exemptions in California." *New England Journal of Medicine* 373 (9): 785–87.

Mnookin, Seth. 2011. *The Panic Virus.* New York: Simon & Schuster.

Mooney, Chris. 2011. "What is motivated reasoning? How does it work? Dan Kahan answers." *Discover Magazine,* May 5. Accessed February 3, 2016. http://blogs.discovermagazine.com/intersection/2011/05/05/what-is-motivated-reasoning-how-does-it-work-dan-kahan-answers/

Mooney, Chris. 2014a. "Study: You can't change an anti-vaxxer's mind." *Mother Jones,* March 3. Accessed May 6, 2015. http://www.motherjones.com/environment/2014/02/vaccine-denial-psychology-backfire-effect

Mooney, Chris. 2014b. "Here's how irrational flu vaccine deniers are." *The Washington Post,* December 8. Accessed May 6, 2015. https://www.washingtonpost.com/news/wonk/wp/2014/12/08/heres-how-irrational-flu-vaccine-deniers-are/

Mulgan, Richard G. 2003. *Holding Power to Account: Accountability in Modern Democracies.* New York: Palgrave Macmillan.

National Conference of State Legislatures. 2016. "States with religious and philosophical exemptions from school immunization requirements." Accessed February 3, 2016. http://www.ncsl.org/research/health/school-immunization-exemption-state-laws.aspx

Nyhan, Brendan. 2015. "Why California's approach to tightening vaccine rules has potential to backfire." *New York Times,* April 14. Accessed November 12, 2015. http://www.nytimes.com/2015/04/15/upshot/why-californias-approach-to-tightening-vaccine-rules-could-backfire.html?abt=0002&abg=0

Nyhan, Brendan, and Jason Reifler. 2015. "Does correcting myths about the flu vaccine work? An experimental evaluation of the effects of corrective information." *Vaccines* 33 (3): 459–64.

Nyhan, Brendan, Jason Reifler, Sean Richey, and Gary L. Freed. 2014. "Effective messages in vaccine promotion: A randomized trial." *Pediatrics* 133 (4): e835–42.

Offit, Paul A., and Susan E. Coffin. 2006. "Communicating science to the public: MMR vaccine and autism." *Vaccine* 22 (1): 1–6.

Omer, Saad B., Daniel A. Salmon, Walter A. Orenstein, Patricia deHart, and Neal Halsey. 2009. "Vaccine refusal, mandatory immunization, and the risks of vaccine-preventable diseases." *New England Journal of Medicine* 360 (19): 1981–88.

Omer, Saad B., Jennifer L. Richards, Michelle Ward, and Robert A. Bednarczyk. 2012. "Vaccination policies and rates of exemption from immunization, 2005–2011." *New England Journal of Medicine* 367 (12): 1170–71.

Oxenham, Simon. 2015. "The backfire effect: When correcting false beliefs has the opposite of the intended effect." *Big Think,* March. Accessed May 11, 2015. http://bigthink.com/neurobonkers/the-backfire-effect-when-correcting-false-beliefs-has-the-opposite-of-the-intended-effect

Page, Clarence. 2015. "A vaxxing dilemma." *The Philadelphia Tribune,* February 13. Accessed May 11, 2015. http://www.phillytrib.com/commentary/a-vaxxing-dilemma/article_2d41d6f4-ece4-5b15-b2a3-571eadc80d65.html

Peeples, Lynne. 2015. "Anti-vaccine haven digs in as measles outbreak hands science crusaders an edge." *The Huffington Post,* March 6. Accessed May 11, 2015 http://www.huffingtonpost.com/2015/03/06/vaccines-exemption-measles_n_6812092.html

Pugh, Tony. 2014. "U.S. faces worst measles outbreak in 20 years." *Toronto Star,* May 29. Accessed February 3, 2016. http://www.thestar.com/news/world/2014/05/29/us_faces_worst_measles_outbreak_in_20_years.html

Rogers, Everett M., James W. Dearing, and Soonbum Chang. 1991. *AIDS in the 1980s: The agenda-setting process for a public issue (Journalism Monographs, No. 126)*. Columbia: Association for Education in Journalism and Mass Communication.

Romm, Cari. 2014. "Vaccine myth-busting can backfire." *The Atlantic*, December 12. Accessed May 7, 2015. http://www.theatlantic.com/health/archive/2014/12/vaccine-myth-busting-can-backfire/383700/

Ropiek, David. "Vaccine exemptions should be harder to get, but don't eliminate them." *Los Angeles Times*, February 9. Accessed January 12, 2106. http://www.latimes.com/opinion/op-ed/la-oe-0210-ropeik-vaccine-exemption-ban-20150210-story.html

Sanders, Lisa. 2015. "Vaccines: The best way to persuade parents is the worst for kids." *Reuters*, February 4. Accessed May 11, 2015. http://blogs.reuters.com/great-debate/2015/02/03/vaccines-the-best-way-to-persuade-parents-is-the-worst-for-kids/

Selbig, Aaron. 2015. "Study shows Michigan's vaccine education plan could backfire." *Interlochen Public Radio*, March 5. Accessed May 11, 2015. http://interlochenpublicradio.org/post/study-shows-michigans-vaccine-education-plan-could-backfire

Singal, Jesse. 2015. "Politicizing the vaccine fight could make things worse." *New York Magazine*, February 5. Accessed May 7, 2015. http://nymag.com/scienceofus/2015/02/dont-politicize-the-vaccination-fight.html#

Stafford, Tom. 2015. "Throwing science at anti-vaxxers just makes them more hardline." *The Conversation*, February 19. Accessed May 13, 2015. http://theconversation.com/throwing-science-at-anti-vaxxers-just-makes-them-more-hardline-37721

Strauss, Mark. 2014. "Debunking flu shot myths makes some less likely to get vaccinated." *iO9*, December 9. Accessed May 7, 2015. http://io9.gizmodo.com/debunking-flu-shot-myths-makes-some-less-likely-to-get-1668803134

Thaler, Richard H., and Cass R. Sunstein. 2008. *Nudge: Improving Decisions about Health Wealth and Happiness*. New York: Penguin Books.

Tremonti, Anna Maria. 2015. "How to convince anti-vaxxers to get their kids vaccinated." *CBC*, February 6. Accessed May 6, 2015. http://www.cbc.ca/radio/thecurrent/feb-6-2015-1.2952341/how-to-convince-anti-vaxxers-to-get-their-kids-vaccinated-1.2952346

Vestal, Christine. 2015. "In states with looser immunization laws, lower rates." *The Pew Charitable Trusts*, February 9. Accessed February 3, 2016. http://www.pewtrusts.org/en/research-and-analysis/blogs/stateline/2015/2/09/in-states-with-looser-immunization-laws-lower-rates

Ward, Stephen J. A. 2011. *Ethics and the Media: An Introduction*. Cambridge: Cambridge University Press, 2011.

Wolfe, Anna. 2015. "How to convince anti-vaxxers to get their kids vaccinated." *Jackson Free Press*, February 18. Accessed August 9, 2015. http://www.jacksonfreepress.com/news/2015/feb/18/anti-vax-paradox-pitting-parental-freedom-against-/

Chapter 8

Values as "Evidence For"

Mental Illness, Medicine, and Policy

Susan C. C. Hawthorne

The question will arise, Is [community care for people with mental illnesses] medically and psychiatrically sound? We can say unhesitatingly that it is medically and psychiatrically sound.

Francis J. Braceland ("Hearings" 1963, 246)

Social reform and humanitarian progress do not necessarily depend on acquisition of scientific knowledge. ... Rather, they may depend on public conviction that a class of people is being wronged.

Joint Commission on Mental Illness and Health (1961, 6)

Weeks before John F. Kennedy's assassination, Congress almost unanimously passed Public Law 88–164, Title II of which is referred to as the "Community Mental Health Centers Construction Act" ("Mental" 1963).[1] Reflecting and furthering multiple social and medical trends, this legislation helped fundamentally reshape the U.S. approach to serious mental illness. By funding construction of local, rather than state, comprehensive mental health facilities, as well as research, the Act greatly decreased prolonged institutionalization of mentally ill people in underfunded, grossly overcrowded, and sometimes abusive state hospitals, moving to a system that relies primarily on care in community-based mental health centers. The legislation enjoyed overwhelming support from the medical community, represented by the American Medical Association, the American Psychiatric Association, the American Psychological Association, national organizations for hospitals, nurses, social workers, occupational therapists, and many other groups.

Fast forward: Passage of the Community Mental Centers Health Act was followed by multiple legal actions protecting the civil rights of people with

133

mental illness. Nevertheless, in the past 50 years, no state or medical system in the United States has established an adequate system of community
care. At present, roughly 350,000 people with serious mental illness are
in prison—10 times more than are in state hospitals (Stettin, Geller et al.
2014, "Callous" 2015). Mental illness contributes to homelessness as well:
approximately 30% of people who are chronically homeless have a mental illness other than substance abuse; an overlapping 50% have a substance abuse
disorder (United States b 2011). What was envisioned, in 1963, as humane
deinstitutionalization has morphed into transinstitutionalization to prison, or,
for many, to frank neglect.

In 1963, Francis J. Braceland, former president of the American Psychiatric
Association, argued before Congress that the evidence favoring the move
away from state hospitals and to community care was medically and psychiatrically sound (see epigraph). Retrospectively, we can doubt this. The medical
community gave its support in a time when psychiatry was transitioning from
psychodynamics to pharmacology, before the initiation of consensus conferences, modern clinical trials, and the evidence-based medicine movement
(Solomon 2015), and at a time when it was acceptable to draw frankly on
social values to bolster medical policy—at least in the absence of definitive
scientific evidence (Joint Commission 1961). The issue is, of course, one of
medical *policy*, as opposed to other types of medical decision-making, such
as choices for individual patients, or calculations of QALYs achieved by an
intervention. Complications of absence of evidence and the distinctiveness of
policy decisions require consideration, given below, but bring us to the central
questions of this chapter: In retrospect, *was* the evidence sound in 1963? And
what can we learn from the 1963 use of evidence, so that we might make
better choices going forward?

It will become clear in Part 1 that even according to the standards of the
time, little medical or scientific evidence existed for the change and that a
convergence of values was a core driver of the bill's passage. However, I will
argue in Part 3, in a broadly pragmatist framework, that policy makers of
the time were not wrong in using values to draw their conclusions: medical
policy recommendations require a rich set of well-defined and well-defended
values as part of the evidence for their justification—values including but
not limited to epistemic values. Parts 3 and 4 refine these points, explaining
that appropriate use of values as evidence requires that those presenting or
assessing arguments be clear what the values are evidence *for*. Further, to use
values legitimately as evidence for medical policy decisions, the values must
be given the same scrutiny needed for all evidence. Part 5 concludes that to
help the millions of mentally ill people ill-served by current medical policy,
we need to come to grips once again with relevant values and facts, so that
we can appropriately guide the next round of reforms.

WHAT WAS THE EVIDENCE?

This section details key evidence from three sources, available to the medical community, that factored into the decisions on the Community Mental Centers Health Act.[2] The first source is information presented to the 1963 Subcommittee of the Committee on Interstate and Foreign Commerce of the House of Representatives ("Hearings" 1963); the second is evidence or trends largely tacit in the testimony, but influential according to other analysts (Dowdall 1996, Paulson 2012); and the third is the "Joint Commission Report" (Joint Commission 1961). This report, mandated by Congress and supported by members of 36 health-care-oriented organizations, was researched and prepared over five years by 45 permanent Commission members and dozens of consultants in various fields.

Evidence Presented to the Subcommittee ("Hearings" 1963)

Over many hours, the 1963 subcommittee reviewing the Community Mental Centers Health Act heard strikingly uniform testimony. For the most part, those testifying spoke as experts, identifying their conclusions rather than their sources of information. Supporting groups included the U.S. Department of Health, Education, and Welfare (HEW), the American Medical Association, and associations representing psychologists, nurses, mental health program directors, rehabilitation professionals, hospital administrators, and state governors; a representative for the pharmaceutical company Smith-Kline & French Laboratories also spoke. Eight themes appeared repeatedly in the testimony, along with a few objections:

1. *Number One Health Problem.* Early in the hearing, HEW head Anthony J. Celebrezze termed mental illness (and mental retardation, the other target of the bill) "national health problems of tragic proportions compounded by years of neglect" (43). Others reinforced this claim with prevalence or severity statistics, for example, stating that 1 in 10 U.S. citizens would at some point need professional help for mental health issues.[3]

2. *Deficiencies of the State Hospitals.* Several cited the figure that, on average, the state hospitals spent only about $4 per day (approximately $31 in 2015 dollars) on their patients (pp. 42, 67, 77, 188, 316). Jack R. Ewalt, Harvard professor and director of the Joint Commission on Mental Illness and Health, described the crowding vividly: "One bed may be so close to the next that you can ride a bicycle across them and never hit the floor" (240). Others testified that committing patients to such institutions constituted a "near barbarian practice of exorcising the mentally sick person to a State institution, miles from home and usually even farther from

recovery" (315), and into the hands of ill-trained "tyrant" doctors, who "are all too often misfits, incompetents, or alcoholics" (381). Based on the prevailing psychodynamic view of mental illness, mental illness was socially conditioned, raising the concern that the abhorrent social conditions of hospitalization actually contributed to chronicity.

3. *Prevention and the Problem of Juvenile Delinquency.* Many speakers cited prevention as a prominent goal of bringing mental health to communities; several specified mitigating the "serious national problem" (330) of juvenile delinquency, along with reducing mental illness in slums more generally. Only one speaker mentioned completed studies of effectiveness of such initiatives (367).

4. *Modern Treatment.* Multiple speakers decried "custodial" models of care, advocating a new model in which rehabilitation would be the norm. They valorized modern medicine and science, which had uncovered a "dramatic advance in our capacity to cope with mental illness" (43) by which "some of the mental illnesses had been conquered, and some others prevented from becoming chronic" (244). The primary advance referred to was the 1950s introduction of "tranquilizing and 'psychic energizing' drugs [chlorpromazine and reserpine], and the shock therapies" (204). However, arguing for the importance of community care, Braceland noted that "some patients leave State hospitals, fail to take their medicines, and where there is nobody to care for them, they neglect themselves and gravitate back to the hospital" (246).

5. *Medical Goals.* The general vision was that people currently institutionalized could, with modern treatment and the shift to community care, achieve "happy and useful lives in the general community." Speakers also aimed at continuity of care: "The patient in [a community mental health] center could move quickly from diagnosis to treatment, from inpatient to outpatient status, from sheltered workshop to industry" (44). Few speakers, however, presented clinical data; most relied on citing the conclusions of the Joint Commission Report. Some compared community programs with state hospitals, citing the endpoint of length of stay. For example, one study found that patients with schizophrenia stayed 11 years on average in state hospitals, while 7 of 10 were discharged from community programs within a year.

6. *Community.* Many testifiers contrasted the state hospital's impersonality and patients' isolation from friends, family, and society with the therapeutic benefits of engagement as well as the greater availability of medical care in the community. Speakers predicted that community members would be supportive because the public had adopted modern, "more enlightened" attitudes to mental illness (98). Over time, speakers claimed, the community members would gain further sense of responsibility for the mentally ill as the patients were brought out of isolation.

7. *Economics.* Expenses for care of the mentally ill were estimated at $2 billion to $3 billion per year. At the time, mentally ill patients occupied half of hospital beds, most (670,000 of 702,000) in 300 large mental institutions.[4] Speakers envisioned economic savings for the hospital system and mental health care system as a whole through greatly reducing hospitalization numbers and length of stay. Several testifiers also observed the costs of mental illness to the military: 25% of armed services rejections were for "mental diseases and deficiencies."

8. *Inadequate Personnel, Future.* Despite the general consensus that the move toward community care was essential, many expressed concern about the lack of adequate professional resources to staff community centers. Yet almost all speakers expressed optimism about remediating the situation quickly through increased training and interest in providing care.

Objections. A few supporters contested details of the bill; only a handful found it misguided. A spokesperson for the psychotherapists of the American Mental Health Foundation advocated group psychotherapy as a low-cost alternative to the proposed medication-centered "modern" treatments envisioned by the majority. A speaker for the Pure Food Association of America, which advocated nutritional approaches to improving mental health and preventing mental illness,[5] argued that benefits of "modern therapy" had been grossly overstated:

> A cult of so-called science writers is greatly responsible for creating an almost failure-proof health image. With the development of the tranquilizers and the drive for ever bigger, annual Government research budgets, the public relations expert has built up a concept of scientific mental health treatment and competence all out of proportion to the ability to deliver. Truly the American public has been sold a bill of goods—that medical science, psychiatry, and modern drugs have eliminated mental health as a serious problem. (385)

A few speakers expressed concern over psychiatric and government overreach in labeling too many as mentally ill. One cited the recently published ideas of anti-psychiatry psychiatrist Thomas Szasz, whose view was that mental illness is not real, but is rather an instrument of social control (Szasz 1961). None of these concerns received much uptake.

Tacit Rationales

Other economic and social factors, largely tacit in the testimony, contributed to swaying opinion in favor of the Community Mental Health Centers Act. Indeed, the push toward deinstitutionalization began as early as the Second World War, when conscientious objectors who served in the state hospitals, along with investigative journalists, made conditions in the hospitals known (Paulson 2012). The popular press also informed the public of the pending legislation ("800 million" 1963). A nascent patients' rights movement, which accelerated after passage of the bill, was also emerging.[6]

The idea that people who have mental illness should receive treatment rather than custodial care was also already familiar. American Psychiatric Association standards for mental hospitals called, as of 1945, for occupational, recreational, and allied therapy; psychotherapy; physio hydrotherapy; and modern organic therapy (Committee 1945, Dowdall 1996). However, most patients received little treatment, and its efficacy was doubted (Joint Commission 1961). The patient population of the state hospitals was also already dropping, as advances in public health reduced the numbers of non-mentally-ill people hospitalized there, such as people infected with tuberculosis.

Social and economic trends decreased the viability of the old model. The state hospital model sequestered the mentally ill and their caregivers from the rest of society on large, residential grounds. But eased transportation access made the model less desirable to most staff, who preferred to live elsewhere, and enlarged population centers engulfed many of the grounds. At the same time, the shift from moral therapy to medical therapy had already reduced the need for expansive and expensive facilities, and, nationwide, the hospitals were aging and in need of expensive updating and repairs (Paulson 2012). Changes in government funding also set the stage: The National Mental Health Act of 1946 and the founding of the National Institute of Mental Health in 1949 had already shifted government support for mental illness research away from state hospitals and toward academic medical centers (Paulson 2012).

New professions and new economic opportunities also shaped the developing consensus. Psychiatrists, relatively poorly paid at state hospitals, could find more lucrative employment within the larger community. Psychiatric nurses, licensed psychiatric aides, research scientists, providers of community care, and drug manufacturers all stood to benefit from the change.

The Joint Commission Report

Action for Mental Health presents the Joint Commission proposal for dealing with the "national emergency" (Joint Commission 1961). Its conclusions rest on the results of 10 government-funded projects that reviewed multiple aspects of the issue: current concepts, epidemiology, attitudes toward mental health, the role of schools and churches, economics, community resources, "manpower," research, and new treatments. In characterizing the magnitude of the problem, state hospital deficiencies, the promise of new medications, and optimism about the potential for community care, the conclusions mostly parallel the testimony before Congress. For example, in "The Tranquilized Hospital," a key section on treatment, the authors review the use of the new major tranquilizer drugs, introduced in 1953 and widely used by 1955. Based on this clinical experience and a few studies, they conclude that the drugs "have revolutionized the management of psychotic patients" (39), offering a *new hope* (40, emphasis in the original). The hope stems, the commission reports, from the direct effect of the drugs on patients and their indirect effects on staff morale: both contribute to staff working more closely and effectively with patients. "The great virtue of the tranquilizers seems to be that they make the patient a more appealing person to all those who must work with him" (53).

Yet the writers express caveats that cast some doubt on the clear testimony before Congress. In its "Summary of Recommendations," the report admits the limited knowledge base, framing its recommendations "in the absence of more specific and definitive scientific evidence of the causes of mental illness" (ix). It cautions that community care will not be appropriate for all: some patients, including the "schizophrenic, involutional, geriatric, and senile" (268), will continue to need prolonged institutional care. In addition, in the section "Why We Reject Psychotics" and elsewhere, the authors express concern that people's attitudes toward the mentally ill are not supportive or enlightened. The writers are also less sanguine than the testifiers about primary prevention efforts, pointing out that more research was needed (242).

The report's writing style is emotionally expressive: Braceland, in a short section expressing his personal disagreement with the document's tax recommendations, describes it as "a wonderful, broad, warm, human document" (330). The report also draws frankly on values to establish its conclusions. For example, the writers motivate their recommendation to greatly increase assistance to "millions of mentally ill" by saying, "To state the case truthfully and place the reality of *what is* side by side with our pretensions as a preponderantly Christian, democratic, humanitarian, scientific, productive society can serve only to embarrass us" (xxvi). That is, because "our" values are these, "we" should feel embarrassed to withhold assistance.

INTERLUDE

At this point, I am expecting my twenty-first-century readers are aware of academic and cultural trends that differentiate accepted academic writing practice in 1963 from accepted practice today. Most twenty-first-century academics would question the assumption that "we" share values and would reject at least some of the values that inflected the 1963 discussions—"Christian, democratic, humanitarian, scientific, productive," or others (xxvi). Many would also find foreign the mash of facts and values evident in the testimony, and consider primitive the 1963 willingness to make sweeping proposals with limited data. Such concerns cast immediate doubt on the reasoning of the Joint Commission, the medical community as a whole, and Congress in their overwhelming support of the Community Mental Health Act. The next two sections recognize that these concerns need careful attention, but argue that, despite these concerns, the 1963 use of values as evidence can point toward an appropriate model—after some significant corrections.

VALUES AS EVIDENCE

This section has two tasks: to demonstrate that values are needed to craft evidence for medical policy decisions and to show that (with caveats) values constitute legitimate evidence for medical policy decisions. Central to my argument is the claim that, in medicine, two different dichotomies are unstable: the (supposed) science/policy dichotomy and the (supposed) fact/value dichotomy. My argument for the former point builds on Miriam Solomon's discussion of the "untidy pluralism" of medical knowledge-making (Solomon 2015); Deweyan reasoning supports the latter claim.[7]

In her recent work on medical knowledge-making, Solomon points out that medical reasoning and decision-making involve multiple, overlapping facets, including basic science, consensus conferences, evidence-based medicine, clinical decision-making in general, Bayesian medical decision-making in particular, narrative medicine, translational medicine, and others (Solomon 2015). These facets vary in their standards, uses, and effectiveness. They differ as well in that they variably prioritize fact-based demonstration over value-laden goals such as innovation, patient preferences, and implementation—or the reverse. Solomon terms this ongoing variability in approach an "untidy pluralism" in the making of medical knowledge. Extending her point, the making of medical policy is also on this continuum, particularly in that it sets the conditions for further knowledge-making and practice;[8] the dichotomy breaks down in the other direction as well, in that additions to knowledge alter policy. Clearly, medical policy making also *differs* from other forms of medical knowledge-making, especially in that the immediate goal of policy making is not to determine facts, as in the basic sciences, but to direct actions. Nonetheless, the dividing lines are not sharp—notice, for example, that clinical decision-making also directs actions, albeit on a smaller scale.

While in a simple form, Hume's point that there is a logical gap between "is" (facts) and "ought" (values) is unassailable (Hume, 1978 [1740]), the supposed fact/value dichotomy is also unstable for multiple reasons. Although it is illogical to conclude the "ought" statement "women *ought* to be beaten" from the "is" statement "women *are* beaten," our language and reasoning are not so simple. One way we argue from fact to value is to find shared values that act like facts. For example, Peter Singer's famous argument in "Famine, Affluence, and Morality" relies on this strategy, offering no argument for the claim "suffering and death from lack of food, water, and medical care are bad" (Singer 1972, 231). In medicine, the logical pivot point between relatively factual claims and more value-laden ones is often a thick ethical term such as "suffering," "care," or "illness."

People often distinguish facts and values metaphysically as well. A common conception is that facts pertain to physically existing entities, while values pertain to human preferences that, like concepts, do not have a physical existence. Epistemological dichotomizing interprets facts as objective, values as subjective. American Pragmatist John Dewey objects against both interpretations that we cannot compartmentalize our world into an "inside" and "outside" (Dewey 1959 [1929]). Experience is interactive and mutually influential: because the rest of the world acts on people and people on the rest of the world, no distinct subjective/objective boundary exists, nor does a fact/ value boundary. Forms of judgment can be roughly distinguished, but they cannot be compartmentalized.

Dewey argues further that—in contrast to logical or mathematical reasoning—most scientific reasoning is practical: the point of reasoning is to choose what to *do* (Dewey, 1998 [1915]). Practical reasoning patches together judgments that vary in their value content. *Practical judgments* direct action toward a particular goal: they orient to the future. *Value judgments* are practical judgments directed toward achieving goods or relieving bads. *Descriptive judgments* orient to the past or present, assessing aspects of the world without acting on it. Descriptive judgments, such as "Ritalin decreases physical activity," equate roughly to "facts"; they usefully contribute to understanding means to achieve goals. Dewey's terms do not simply rename facts and values, however; they instead direct attention to ways facts and values are alike. Descriptive, practical, and value judgments are all *judgments*, and all necessary for practical reasoning. All are hypotheses, not determinate. Very significantly, one must have reasons—grounded in real-world consequences—for any judgments, including value judgments.[9]

In medical science, fact and value overlap for other reasons as well.[10] Briefly, this is because value-laden medical interests in human health and disease imbue all aspects of medical science—questions, language, operationalizations, interpretation, and implementation—so that cleanly disentangling facts from interests, preferences, and goals is not possible. That is, all facts in medical science are value laden. When people use medical science clinically or for policy, they import embedded values to their decisions, often adding new values.

Finally, in both clinical and medical policy decisions, values have a motivating role.[11] Medical or humanitarian impulses, respect for autonomy, and the like provide impetus toward certain choices and away from others. Values weight the scales—provide evidence—for deciding among treatments or adopting policy.

In addition to general fact/value overlap, "evidence" is *always* value laden to some extent because "evidence" is always "evidence *for*" something: even if there were a pure fact (unlikely in medicine, as I argue above), using that fact as "evidence" frames it for a particular purpose. For example, consider disagreements over what constitutes effective care for psychiatric illnesses. In 1963, the medical goal was to return people who had mental illnesses to social functioning, by which was meant a stable place in family, community, and work. A disability rights activist, however, might argue that such visions of social integration express ways of being that some people reject or to which they are not suited—and that asocial, eccentric, or socially marginal lives are worthy lives. For this activist, demonstrating that people can function at work might not constitute evidence for effective treatment; conversely, demonstration of contented, asocial lives might not constitute evidence for effectiveness in the eyes of 1963s therapists. More generally, because values help shape what counts as "evidence for," only some facets of knowledge, observations, possibilities, etc. are considered relevant in collecting (in research) and selecting (for presentation) evidence for or against a particular endpoint. Almost inevitably, there is an imbalance in which facets are valued—and therefore collected or selected—and which are not, pervading the evidence (for) with the values involved.

Recognizing that values necessarily permeate medical policy decisions leaves a question: Is the use of values epistemically legitimate? The answer to this question depends not on *whether* values are used as evidence—they must be—but on *how* they are used. Recognizing values as evidence has the distinct advantage that it draws attention to a crucial step: when values act as evidence for an action, the values need to be defensible, just as does the other "evidence for" (see endnote 9).

Despite the caveat that values need defense, the proposal that values can serve as evidence raises well-warranted concerns about manipulation of science by values and of policy by junk science marred by values. Motivated by such concerns, Heather Douglas, among others, disagrees that values and evidence should be "on a par," either in interpreting evidence within scientific communities or in using science to guide policy decisions (Douglas 2009, 149). Consider first Douglas's prescription for use of values within science. Values may, she argues, legitimately play an indirect role in interpreting evidence, but they should never play a direct role. For example, in judging whether or not the evidence shows that a chemical used in some sunscreens is carcinogenic, a scientist should not consider her brand preference as evidence that the sunscreen is safe. Weighing the evidence in light of that preference would be a direct use of values and illegitimate. In legitimate indirect use, the scientist necessarily uses values to determine the significance of uncertainty in the evidence: in the sunscreen example, she would weigh the risk an incorrect judgment would pose to human health and the environment.

Extrapolating the legitimacy of indirect value use and the illegitimacy of direct use to risk assessment for policy decisions, Douglas writes, "The direct role for values that place them on a par with evidence would allow scientists to ignore evidence because they prefer a different outcome for a risk assessment" (149). The intent is that neither science nor policy be subsumed by values. Note, however, an assumption embedded in this line of thought: the idea that using values as evidence allows a scientist (or anyone) to ignore evidence. I have already noted that values direct attention to particular evidence and that within a value system, evidence may not be relevant evidence (for). However, *ignoring* evidence is not the same as *failing to see* its relevance. The former is dogmatism, not proper use of values (Anderson 2004); the latter, given openness to other points of view, is correctable.

In addition, while I agree that values are *not* on a par with other evidence in contexts such as mathematical reasoning or simple claims backed by clear observation or experiment, they are a necessary component of "evidence for" in contexts such as designing mental illness classification systems, establishing efficacy criteria, and determining courses of action. In situations of crisis, particularly—the mistreatment of mentally ill people, climate change—values play very significant evidentiary and motivational roles.

Daniel Hicks argues for an expanded role for values in science (Hicks 2014). He helpfully identifies various strategies in earlier attempts to describe legitimate influences of values in science. One is Douglas's model, which I have already argued against. Another allows use of values in ways that promote truth, allowing use of epistemic values and eschewing nonepistemic values. Hicks argues that this strategy inadequately incorporates values because it lexically prioritizes truth, automatically downgrading the role of many values. He also points out the practical concern that truth may matter to scientists, but does not matter (at least not in the same way) to many non-scientists. The difference in perspective is problematic because shared commitments are needed for collaborative projects. To solve this problem, Hicks prioritizes the project and suggests that collaborators determine which values contribute to its success. If certain values do not contribute, or are inimical, they should be overridden by other values. Truth will often be one goal of the project, but other value-laden goals—such as human health or gender equality—can take priority.[12]

This account shares commonalities with the view I am developing here. However, my argument refines Hicks's perspective in two related ways. First, it goes further in considering values as a form of evidence, thereby demanding that values be justified. Second, Hicks's account is insufficiently critical of values involved, including those that define the project in question: values are not all equally defensible, and they need to be challenged and debated.

To summarize: uncontroversially, values identify, shape, and prioritize needs and goals. More controversially, within medical sciences (at least—I have not taken on a wider argument here), values inevitably become deeply entangled in factual evidence in the process of gathering and interpreting data. In common practice, the entanglement gives them a role as evidence by default. But the default position allows them to be engaged badly, risking epistemic or moral violence to the evidence. In a second role, values provide evidence for action, serving on their own as a rationale and as motivator. Values are often used unthinkingly in this role as well, again inviting controversy over their use in policy decisions.

These considerations lead to two main caveats about the use of values as evidence: First, even values that are well justified by prior observation and use are rarely sufficient evidence for policy. However, with careful consideration they can work in tandem with, rather than override, more clearly empirical information. Second, values are as contestable as any evidence: To be used well, their role in evidence presented must be made explicit, so that they can be challenged and debated.

USING VALUES AS EVIDENCE: RETROSPECTIVE EVALUATION

This section reconsiders the 1963 passage of the Community Mental Health Centers Act in light of the preceding argument. Did the medical community have the evidence to endorse the Community Mental Centers Health Act? I suggest that this question needs more precision, in order to consider exactly what people of the time had evidence *for*. The discussion also revisits the necessary and legitimate role of values in the decisions.

The medical testimony set goals of offering "modern treatment," returning mentally ill people to social functioning, preventing mental illness, and providing continuity of care. The evidence that moving to community care could achieve any of these goals was decidedly weak, as evidenced by expressed caveats about "manpower," gaps in research, and inexperience with community care. Further, although chlorpromazine and reserpine provided new options for treatment, experts had less than 10 years of experience with these medications—and that was primarily within controlled institutions and a few experimental, hence relatively controlled, community trials. In short, the evidence for the efficacy of "modern therapy" was limited—whatever the definition of "efficacy" might have been—and evidence for its real-world effectiveness was even more so.

What, then, overwhelmingly convinced the medical community that the move to community care was appropriate, so that Braceland could say truthfully, *"We* can say unhesitatingly that it is medically and psychiatrically sound" ("Hearings" 1963, 246; emphasis added)? Admittedly, answering this question requires some speculation. The gap between the evidence and the conclusion "support the Act" might have been filled by adherence to the then-reigning psychodynamic model of care, which hypothesized psychogenic causes for most mental illnesses ("Hearings" 1963, American Psychiatric Association 1968). In this model, hospitalization itself was in principle toxic and supportive community in principle salutary—in short, the gap could be filled by a priori reasoning. Alternatively, the gap might have been filled by ignoring evidence against the change—by cherry-picking evidence to match their predetermined preference and ignoring the rest, the medical community would have reached the predetermined conclusion. Such direct use of values is what concerns Douglas. I think that the most likely gap-filler, however, was that the medical community used values—some specified, some not—as a form of evidence. The combination of their "Christian, democratic, humanitarian, scientific, productive" values and assorted medical, practical, and economic values provided rationales that converged on endorsement (Joint Commission 1961, xxvi).

So, was it epistemically permissible to fill the gap between evidence and conclusion by using values as evidence? Yes (qualified), and No. *No* because it was an error to mistake the multiple values embedded in the stated goals for evidence of effectiveness of the proposed community care. While the goals for community care were reasonable, given the values employed, there was little evidence that the policy would actually achieve the goals. A qualified *Yes* because the values did provide *some* support for the change, because the medical community was also addressing another question: Should we move mentally ill people out of the state hospitals?

The evidence for moving people out of the state hospitals was overwhelming[13]—and recognition that it was so depended on values. Spending $4 a day on care, cramming people into wards, physical restraints, and so forth are *bad* only through the lens of specific values—these features could be good if one's values prioritized economy. The medical community's embrace of Christian, humanitarian, and democratic values,[14] gave it solid reasons to recognize the state hospitals as dysfunctional and unethical, and the resulting decision to shift the emphasis in care for the mentally ill provided a new framework, making a new—and in some ways better[15]—practice of medicine possible.[16]

Refinements

I have already emphasized the need for values-as-evidence to be thoroughly vetted and debated. The risks of using values as evidence are abundantly clear, bringing to mind pseudoscientific conclusions that rely on sexist, heterosexist, racist, capitalist, nationalist (etc.) assumptions. In this section I further qualify my "Yes" to legitimate use of values as evidence, first considering situations of ignorance compared with situations of deeper experience, and second reconsidering whether any and all values can be used as evidence.

In 1963, "modern treatment" was more a wish than a verified or even defined set of approaches, while knowledge of the ineffectiveness and inhumanity of the old system was robust. In contrast to my analysis of the reasoning, then, one could argue that the values were irrelevant: It was factual evidence of ineffectiveness that made dismantling the old system a good decision, and lack of evidence for effectiveness made implementing the new one hazardous. I favor a different interpretation: The values in both cases provided *the* reason—and an ample reason—to act. However, given the humanitarian values-as-evidence, the need to dismantle (or radically change) the state hospital system was obvious even with *less* factual evidence available—that is, facts and values together provided a relatively clear path forward. In contrast, while the humanitarian need for a new system was similarly obvious, *more* factual evidence was needed to plan an untried system. In these cases—and in all other cases of medical policymaking—neither the factual evidence nor the values-as-evidence can be set aside; both are necessary on some sliding and negotiable scale. In times of crisis, values-as-evidence may legitimately lead, even with relative ignorance. For example, humanitarian use of novel treatments draws on this recognition. With greater knowledge available, or with less imminent need, factual evidence rightly receives additional weight.

The idea that some values can successfully guide science while others cannot has received attention particularly within discussions of socially responsible science (Kourany 2010, Kitcher 2011, Intemann 2015). As tempting as it is for me to designate feminist, humanitarian, or democratic values as guides, I think that the better approach is to teach and establish an overarching value of nondogmatism, and rule out only its contrasts, such as intransigence, inflexibility, and infallibilism. Doing so implies the need to include varying viewpoints and a process that can support careful identification, debate, and defense of values (Longino 1990). No such investigation took place in 1963—rather, shared values were assumed and not subjected to analysis. In today's politics, in contrast, it is common to recognize disagreement over values but to be unwilling to hear evidence against one's own perspective. The result is polarization and stasis on crucial issues, including mental health policy.

My conclusion, then, is that well-vetted values serve necessarily and legitimately as evidence for action, as long as they are appropriately weighted to reflect the urgency of a need *and* the quality and quantity of factual evidence. This is not foreign territory in medical reasoning: values provide evidence in choices for individual patients as well.

USING VALUES AS EVIDENCE: TODAY'S DECISIONS

Today's circumstances leave millions of mentally ill people without access to standard psychiatric care, without sustainable alternatives, and suffering the neglect of homelessness or imprisonment. At this writing, U.S. citizens are increasingly recognizing the poor options for people with serious mental illness such as the schizophrenias, bipolar depression, and intractable major depression. However, today's advocates for mentally ill people disagree on what the change should be. A recent editorial in *JAMA* calls for bringing back asylums (Sisti, Segal et al. 2015). A national consensus statement issued by the Substance Abuse and Mental Health Services Administration (SAMHSA), urges that the mental health system should be reformed by focusing on recovery[17] from mental illness (United States a 2006). The Treatment Advocacy Center and a bill in Congress call for expanded civil commitments to outpatient treatment, such as mandated medication (Stettin, Geller et al. 2014, Ornstein 2015). The anti-psychiatry movement advocates freedom of choice (Oaks 2011).

The neglect and mistreatment of those who are seriously mentally ill presents a humanitarian emergency and demands action. Basic ethical values and understanding of current circumstances are enough to see that. But to decide among the options proposed above—or others—we need two additional forms of evidence: empirical evidence for effectiveness, and an explicit and thoroughly debated system of values that can be used to articulate what the goals are and what the standards of effectiveness should be. Crucially, in another update to the 1963 approach, the values and experiences of mentally ill individuals need to be included in the decision-making process. The meta-value of nondogmatism requires that we be open to solutions outside the current boxes—perhaps even to radical solutions involving novel goals, invigorated acceptance of people who have mental illness, new professionalisms, or alternative therapies. For appropriate and sustainable change, we need a coalition akin to 1963's—but this time, a coalition with the evidence of vetted values and factual evidence to match.

ACKNOWLEDGMENTS

Thank you to St. Catherine University students in my Fall 2015 Global Search for Justice class for their ideas and inspiration, and Dan Hicks, Amy Ihlan, and Robyn Bluhm for insightful comments on earlier drafts.

NOTES

1. Title I provided resources for care of the "mentally retarded" and research on "mental retardation."
2. This is obviously not a full historical analysis, which would require much more archival research.
3. Interestingly for today's discussions, this figure has expanded to nearly 1 in 2 (Reeves et al. 2011).
4. "Senile dementia" (such as today's Alzheimer's disease) was included among the mental illnesses at the time.
5. Recent success against phenylketonuria was a focus of his testimony.
6. Physician and lawyer Morton Birnbaum argued in 1960 for the idea that people with mental illness were "entitled to treatment" (Paulson 2012).
7. The argument against the fact/value dichotomy paraphrases my argument in an earlier publication (Hawthorne 2014).
8. For example, the Community Mental Health Centers Act established research priorities and funding as well as changing how medical care would be provided.
9. Exactly how value judgments (or values in general) can be defended—exactly like observed facts? Like theorems?—is another issue I cannot take up here. Gilbert Harman argues that the defense of values differs from both mathematical reasoning and observational demonstration (Harman 1988).
10. My full argument for this claim draws on the work of Elizabeth Anderson, John Dewey, and Hilary Putnam (Anderson 2004, Dewey 1959 [1929], Dewey 1998 [1915], Putnam 2002).
11. I cannot here defend or tightly characterize this claim. It is a generally moral realist position (Sayre-McCord 1988).
12. Thank you to Dan Hicks for this way of phrasing the point.
13. An alternative would be reforming the hospitals, but the economics of that choice was also overwhelming (see the section "Tacit Rationales").
14. To make the claim that these provide solid reasons, I take a charitable interpretation of what Christian, humanitarian, and democratic values consist in.
15. Discussing the pros and cons of today's psychiatric approaches is beyond the scope of this essay.
16. Points parallel to those for the medical community can be made for Congress. While Congress might have altered portions of the Community Mental Health Act, the basic question they confronted was whether they had evidence for its passage. The hearing transcript suggests that the Congressional committee accepted the medical community's near-uniform expert testimony with little skepticism. Nonmedical testifiers

emphasized solving the problem of juvenile delinquency, reducing mental illness in slums, and reducing expenses. However, as with the medical testifiers, they presented little factual evidence that the new system would meet the goals, and again, the committee challenged little in their arguments. Congress thus made the double mistake of confusing the value-laden goals for adequate evidence of efficacy, and trusting expert witnesses too much. Thus, like the medical community, Congress had inadequate evidence for the community care's effectiveness. However, as for the medical community, values provided powerful evidence that people should be moved from the state hospitals.

17. Recovery, as opposed to chronicity, disability, etc.

REFERENCES

"$800 Million Due for Mental Ills: U.S. Reveals Projected Cost of Five-Year Program." *The New York Times* April 10, 1963, Late City Edition ed., sec. A. Print.

American Psychiatric Association, Committee on Nomenclature and Statistics. 1968. *Diagnostic and Statistical Manual of Mental Disorders,* 2nd Ed. Washington, DC: American Psychiatric Association.

Anderson, Elizabeth. 2004. "Uses of Value Judgments in Science: A General Argument, with Lessons from a Case Study of Feminist Research on Divorce." *Hypatia* 19.1:1–24.

"Callous and Cruel: Use of Force against Inmates with Mental Disabilities in Us Jails and Prisons." Human Rights Watch. Web. August 25, 2015.

Committee on Psychiatric Standards and Policies of the American Psychiatric Association. 1945. "Standards for Psychiatric Hospitals and Out-Patient Clinics Approved by the American Psychiatric Association (1945–1946)." *American Journal of Psychiatry* 102.2:264–69.

Dewey, John. 1959 [1929]. *Experience and Nature,* 2nd ed. New York: Dover Publications, Inc.

Dewey, John. 1998 [1915]. "The Logic of Judgments of Practice." *The Essential Dewey: Volume 2: Ethics, Logic, Psychology.* Ed. Hickman, Larry A., Thomas M. Alexander. Bloomington and Indianapolis: Indiana University Press. 236–71.

Douglas, Heather E. 2009. *Science, Policy, and the Value-Free Ideal.* Pittsburgh, PA: University of Pittsburgh Press.

Dowdall, George W. 1996. *The Eclipse of the State Mental Hospital: Policy, Stigma, and Organization.* Suny Series in the Sociology of Work. Albany: State University of New York Press.

Harman, Gilbert. 1988. "Ethics and Observation." *Essays on Moral Realism.* Ed. Sayre-McCord, Geoffrey. Ithaca, NY: Cornell University Press. 119–24.

Hawthorne, Susan. 2014. *Accidental Intolerance: How We Stigmatize ADHD and How We Can Stop.* New York: Oxford University Press.

Hearings before a Subcommittee of the Committee on Interstate and Foreign Commerce. Congress 88, First Session, House of Representatives. Washington, DC: U. S. Government Printing Office, 1963. Print.

Hicks, Daniel J. 2014. "A New Direction for Science and Values." *Synthese* 191.14:3271–95.

Intemann, Kristen. 2015. "Distinguishing between Legitimate and Illegitimate Values in Climate Modeling." *European Journal for Philosophy of Science* 5:217–32.

Joint Commission on Mental Illness and Health. 1961. *Action for Mental Health; Final Report, 1961*. New York: Basic Books.

Kitcher, Philip. 2011. *Science in a Democratic Society*. Amherst, NY: Prometheus Books.

Kourany, Janet A. 2010. *Philosophy of Science after Feminism*. New York City: Oxford University Press.

Longino, Helen E. 1990. *Science as Social Knowledge: Values and Objectivity in Scientific Inquiry*. Princeton, NJ: Princeton University Press.

Mental Retardation Facilities and Community Mental Health Centers Construction Act of 1963. Pub. L. 88–164. 71 Stat. 717. 31 Oct. 1963.

Oaks, David W. 2011. "The Moral Imperative for Dialogue with Organizations of Survivors of Coerced Psychiatric Human Rights Violations." *Coercive Treatment in Psychiatry: Clinical, Legal and Ethical Aspects*. Eds. Kallert, Thomas W., Juan E. Mezzich, and John Monahan. Chichester, West Sussex, UK; Hoboken, NJ: Wiley-Blackwell. 187–211.

Ornstein, Norman J. "How to Help Save the Mentally Ill from Themselves." *The New York Times* November 17 2015.

Paulson, George W. 2012. *Closing the Asylums: Causes and Consequences of the Deinstitutionalization Movement*. Jefferson, NC: McFarland & Co.

Putnam, Hilary. 2002. *The Collapse of the Fact/Value Dichotomy and Other Essays*. Cambridge, MA: Harvard University Press.

Reeves, William C., Tara W. Strine, Laura A. Pratt, William Thompson, Indu Ahluwalia, Satvinder S. Dhingra, Lea R. McKnight-Eily, et al. 2011. "Mental Illness Surveillance among Adults in the United States." *Morbidity and Mortality Weekly Report (supplements)* 60.3:1–32.

Sayre-McCord, Geoffrey. 1988. *Essays on Moral Realism*. Ithaca, NY: Cornell University Press.

Singer, Peter. 1972. "Famine, Affluence, and Morality." *Philosophy and Public Affairs* 1.3:229–43.

Sisti, Dominic. A., Andrea G. Segal, and Ezekiel J. Emanuel. 2015. "Improving Long-Term Psychiatric Care: Bring Back the Asylum." *JAMA* 313.3:243–4.

Solomon, Miriam. 2015. *Making Medical Knowledge*. First edition. Oxford: Oxford University Press.

Stettin, Brian, et al. 2014. *Mental Health Commitment Laws: A Survey of the States*. Arlington, VA. Web. 1 Apr. 2016.

Szasz, Thomas S. 1961. *The Myth of Mental Illness: Foundations of a Theory of Personal Conduct*. New York: Hoeber-Harper.

United States a. Department of Health and Human Services. 2006. "National Consensus Statement on Mental Health Recovery." Ed. Administration, Substance Abuse and Mental Health Services: Substance Abuse and Mental Health Services Administration. Web, http://store.samhsa.gov. 1 Apr 2016.

United States b. 2011. Substance Abuse and Mental Health Services Administration. "Current Statistics on the Prevalence and Characteristics of People Experiencing Homelessness in the United States." Web. 2015 July 1.

Chapter 9

RCTs and EBM as Gold Standards

A Trojan Horse for Biomedicine?

Keekok Lee

Randomized controlled trials (RCTs) and evidence-based medicine (EBM) are held up as gold standards for evaluating the effectiveness and efficacy of medical interventions today. This chapter attempts to demonstrate the ironic potentiality of these gold standards to undermine biomedicine itself, as their implications appear to be incompatible with the foundations of that medicine in three related ways: (1) its underlying dualism; (2) its linear model of disease (as an entity) in terms of one cause, one effect; (3) its fundamental presupposition that all patients are homogenous and uniform (unless differences deemed to be relevant are identified and controlled for). Consider (1) in conjunction with the placebo response, consistently seen in RCTs but consistently ignored to date. This head-in-the-sand approach is not sustainable. However, admitting the relevance of the placebo response to understanding the relationship between treatment and illness would ultimately result in questioning dualism, which privileges body over mind in the construction of the monogenic conception of disease, and the linear mono-causal model, which holds between cause and effect in biomedicine.

As EBM rests on RCTs, whatever flaws infect the one would therefore also infect the other. Furthermore, EBM presents another problem, as it has in practice provoked another source of authority in medicine, namely the field clinician, to pose a challenge to the statistical (objectivized) findings of EBM. This brings out the conflict between two very different types of epistemological contexts—on the one hand, experimental data distilled from RCTs by EBM scientists using statistical tools of one kind or another, and on the other, the subjective clinical judgment based on detailed knowledge of individual patients. The challenge of the field clinician is not simply threatening as a rival "political" authority to RCTs and EBM; it could also be threatening (once its implications are fully appreciated via (1) and (2) above) to the theoretical underpinnings of biomedicine itself. It turns out that biomedicine is premised on the homogeneity and uniformity of patients, as otherwise, there could be no repeatability of experiments, nor would it make sense for RCTs and EBM to argue that they serve as gold standards for the effectiveness and efficacy of treating a particular disease with a particular drug. In other words, the notion of "the patient" is an idealized/abstract construction that bears little or no resemblance to actual patients whom the field clinicians have under their care. For doctors at the front line of patient care, the objectivized data yielded by the two gold standards are at best of limited relevance in the clinical context. The pressing fundamental question facing them is, what is the most relevant and best treatment for this particular patient whose long list of specificities could not possibly have been taken into account through the randomization procedure, and not the question, what is the best/most effective intervention for any idealized/generalized/homogenized patient diagnosed as suffering from a particular disease entity called X.

RCTs AND EBM

As a hallmark of scientificity, canonically, RCTs must satisfy not merely single but also double blinding with both patients and the medical staff administering the medicine not knowing who are getting the real and who the fake treatment. Furthermore, which patient would be assigned to the experimental arm (the group taking the treatment under test) or the control arm (the group taking the fake/alternative treatment) must be decided by randomization, preferably today by means of an algorithm, controlled by a computer. These three conditions may be said to constitute the gold standard in its Platonic form.

Unfortunately, in real life, not all forms of medical intervention are equally susceptible to successful faking to control for the placebo effect—at one end, pills are the easiest, and at the other end, major surgery would not be (although some RCTs have successfully used sham surgery). Ethical constraints may also operate to restrain ingenuity in setting up fake interventions. As a result, medical research in many contexts may even use the open-label method instead, dispensing with blinding altogether. This then leaves randomization to carry the burden of upholding the RCT gold standard, a matter to which we shall return a little later.

Evans et al. (2011) tolerate further paring down of the gold standard requirements by recognizing that RCTs are invoked not in all contexts of testing but only in some—those exempt, for instance, include new treatments producing "dramatic results." RCTs are considered to be peculiarly relevant in cases where treatment produces moderate effects (e.g., concerning the relative merits of different types of hip replacement joints or whether a broken leg should be encased in plaster or not).

Since the 1990s, another notion has been appended to the gold standard of RCTs, that of EBM. Both of these are adopted by the Cochrane Collaboration (CC), which is dedicated to organizing and assessing medical research information in a systematic way so that intelligent and reliable choices can be made by patients, health professionals, and policy makers whether in or outside government. The CC's logo encapsulates excellently its intellectual message, what it means by a meta-analysis of RCTs (Kristiansen and Mooney 2004). However, as we shall see, both RCTs and EBM have been subjected to criticisms and that the flaws identified in the former would affect the latter.

Some critics of RCTs (e.g., Cartwright 2007; Kristiansen and Mooney 2004; Pedersen 2004) argue that RCTs fall down badly on what they call external validity (what they mean by this concept will be set out below) but not on internal validity, meaning that RCTs can be conducted so that they have no internal/design flaws. I shall argue that this distinction breaks down, as these are intertwined, being aspects of what may be called scientific validity; hence such a distinction is not very useful.

Most RCTs fall down badly on external validity, in part because of the extraordinary power wielded by the pharmaceutical industry, reflecting the undemanding attitude of governmental/regulatory agencies toward it in enforcing strict methodological requirements of scientific propriety. To remedy the most critical epistemological defect, critics such as Angell (2004), Chalmers (2006), Evans et al. (2011), and Goldacre (2012) argue that all trials must be registered with their protocols attached; all their results, whether negative or positive, must be published in full for independent critical assessment, and the financial interests and backing for such trials must be declared.

The laissez-faire attitude permitted by regulatory agencies means that Big Pharma can do what they please, which is not in keeping with what scientificity demands, as evidence shows that the type and source of funding can influence the quality and the objective assessment of test results (see Angell 2004). This, in turn, means that questioning the "external validity" of RCTs immediately leads to questioning their "internal validity" as well—for instance, a properly mounted RCT (which satisfies the criteria of internal validity) but which yields a negative outcome for the new drug tested is very likely to be suppressed by the pharmaceutical company sponsoring the test. To see this claim better, let us return to RCTs relying on randomization as a technical statistical tool to carry the burden of the gold standards. The researchers set out to control for factors/variables known to make a difference to the outcome of a trial. Hence, they must make sure that the experimental and control arms are the same except in respect of one variable only (following the logic of Mill's method of difference—Mill, 1843/1974: Book III, chapters 8–10).

However, the state of scientific knowledge (at any one time) does not permit us to identify and control for all possible relevant factors, which could make a difference to the test outcome. Randomization is, then, expected to take care of all extraneous factors not yet known or controllable but which could be relevant. Can it bear this heavy weight?

In some cases it may but not in all. A study, say in the USA, may be impeccably run according to the gold standard of RCTs (including randomization), but it happens to have been undertaken in a geographical location that contains predominantly white females; its results, therefore, would certainly hold for "white females" of a certain age, but can one with confidence extrapolate them to U.S. females, say, of Caribbean ancestry, or indeed, to similarly aged females worldwide? One would be methodologically foolish to do so without hesitation. Hence, satisfying the gold standard of randomization (constituting a key component of internal validity) may only be a necessary, but not a necessary and sufficient condition, for "scientifically reliable" data relevant to medical decision-making. Big Pharma have to chase profits; they will develop and test drugs for what they consider to be profitable markets. Furthermore, they have to cut down on production costs; hence, it would not be in their interest to test a drug in as many countries in the world as there are people with different cultural/ethnic and genetic diversities to ascertain what the outcome could be for such populations. Drug trials as such inevitably mask potential incompatibilities between internal and external validity.

To meet this kind of criticism, EBM could conceivably come to the rescue of RCTs by maintaining that systematic reviews of all properly conducted RCTs worldwide would remedy the flaw just characterized. However, in reality, conditions in many parts of the world are not amenable to the running of RCTs, and even if some proper RCTs were run, the results would probably be published in a language and in journals not available to medical researchers in the First World ("democratic, advanced/mature economies").

This rescue offer by EBM may not be overhelpful, as it itself has been criticized for adhering to a dogmatic hierarchy of which (logically) RCTs occupy the first rung, thereby downgrading other studies (in a descending order), namely, cohort studies, case-control studies, case series, case reports—Thelle (2004) holds that RCTs and EBM have not provided good evidence, or, indeed, even evidence at all for advocating such a hierarchy. Furthermore, Kaptchuk has argued that whatever evidence RCTs themselves proffer is rendered tautologically true, as "(RCT) authenticates itself: 'the truth is what we find out in such and such a way. We recognize it as truth because of how we find it out. And how do we know that the method is good? Because it gets at the truth'" (2001, 543). Kaptchuk adds that it is assumed that "any differences between masked RCTs and other research methods are due to deficiencies in the less stringent method. The ideal masked RCT is a priori considered a perfect tool and always innocent of any contribution to distortions from 'reality.'"

Yet other critics have focused on the notion of "meta-analysis" itself, considered by its exponents to constitute the core of the truth in EBM. Kristiansen and Mooney (2004) and Rasmussen (2004) point out that different meta-analyses of the same subject may reach different conclusions even when the studies included are about the same, as independent readers and authors of these reviews rate them differently, with the latter being more positive about the treatment under review than the independent readers. Moreover, Kristiansen and Mooney (2004) point out that the predictive power of meta-analysis is not overwhelming as it appears to predict the results of subsequent large clinical trials only in two out of three cases.

It follows from the above presentation of the arguments that it makes little or no methodological sense to keep separate (a) the scientificity (or otherwise) of RCTs and that of EBM as the two are interlinked and (b) the notion of internal validity from that of external validity of (RCTs and EBM) as these two, too, are interlinked. One needs to grasp that in both domains, some critical flaws in the one can be traced to flaws in the other and vice versa.

AXIOM OF HOMOGENEITY

We now turn to explore in further detail how these flaws are interrelated as aspects of scientific validity. We need to grasp the framework presupposed by RCTs. This framework is equivalent to setting up an axiomatic system. What could be the basic axiom of RCTs? To answer this, let us set out in the briefest, crudest outline possible how RCTs are conducted. First, patients are identified as suffering from a specified disease for which the treatment under test is intended to be effective. Second, out of this group, only those who are prepared to give consent to being part of the trial would become participants. Third, once this group has been identified as set out above, such participant-patients would be randomly assigned either to the experimental or to the control arm of the test. Exponents of randomization say its raison d'être is to prevent bias in the characteristics of participant-patients, which could make a difference to the test outcome, so that any difference in outcomes between the two arms can be explained only in terms of the difference in treatment. Narrowing down the explanation of the test outcome in terms of only one variable (that is, the difference between real treatment or the fake treatment, the placebo) is entirely, as we have already observed, within the logic of Mill's method of difference. However, is it correct to assume that randomization is powerful enough as a tool to overcome all problems of bias including selection bias? It can cope with allocation bias, say, in respect of the severity or otherwise of the illness of the participant-patients—those members of the medical staff, who subconsciously want the new drug to be seen to be effective, might allocate the less ill participant-patients to the experimental group and the more ill to the control arm. Although this kind of bias could be taken care of under randomization, can all differences in characteristics between participant-patients be ironed away by randomization?

As this issue is not explicitly addressed by RCTs, it must be assumed either that their proponents do not recognize that such a problem exists or that the problem can be conjured away by assuming that in respect of these other differences, they are deemed not to exist for the purpose of the RCTs. It is this *deeming* that appears then to be itself problematic. Some of the criticisms directed at RCTs and EBM earlier set out may be read as an implicit questioning of this very presupposition. In other words, participant-patients are deemed to be homogenous in respect of these other features or aspects of their existence, such as their respective genetic inheritances, the environments they inhabit, their diets, their lifestyles, and so on. For this reason, this author proposes that the axiom of RCTs be called *the axiom of homogeneity*, which deems all patient-participants are uniform (enough) in all aspects other than those that are directly taken into account by the trial itself. It is this deemed homogeneity that can make sense of the elevation of randomization (in the allocation of participant-patients to either the experimental arm or the control arm of the trial) to the status of the key, indeed, the only, strategic/analytical tool required, to get rid of all differences that could give rise to bias save that of the treatment itself. Without *the axiom of homogeneity*, randomization would only make limited sense by preventing allocation bias only. It is also this deemed homogeneity that makes the results of RCTs, at best, of limited use and at worst of no use at all in the light of the distinction made in the next section between "statistical relevance" and "clinical relevance."

STATISTICAL AND CLINICAL RELEVANCE

The test results of RCTs are cast in statistical terms. First, for the results to be of any use, there must exist a statistically significant difference between the two arms of the test, commonly referred to as the p value (see Rumsay 2016 for a brief account). Second, there is a concept called confidence intervals (CIs), which refer to the estimates of what might happen if the treatment were to be given to the entire population of interest (that is, the entire population with that illness for which the treatment has been tested). The physician is then expected to read clinical significance into the p value and the CIs—see Akobeng (2005). Sometimes, the notion of NNT is also invoked, as being more user friendly than the p value and the CIs. NNT (number needed to treat) as a popular measure of the effectiveness of a medical intervention; NNTs are calculated from systematic reviews of RCTs; see Moore 2009.) However, the p value, the CIs per se, do not bear direct relation to the task facing the field clinician, namely, of working out what relevance the results of an RCT could have for the individual patients under their charge—for want of a better term, one may call this "clinical relevance." From such a standpoint, RCT test results per se do not bear any direct relation to clinical relevance, which is what doctors and their patients are primarily interested in. Clinical relevance does not coincide with "statistical relevance"—the former is what the patient and the doctor in charge of the patient are interested in, the latter with RCT test results. (See Kalanthini 2016 for admitting the difference between these two criteria[1] and for a moving personal appreciation of the difference both as doctor and as patient.) That is to say, on the one hand, should there be a positive result in favor of a treatment, even if that difference may not be considered to be statistically significant, the patient and doctor could be interested. In other words, it looks as if that what is of clinical relevance and decisive for the patient and doctor is not what RCTs and EBM directly recommend. On the other hand, even if the "best" available evidence by way of RCTs and EBM confirm that a particular drug is statistically effective/statistically relevant, say, to asthma treatment, this does not mean that all asthma patients, including this particular patient under the care of a particular doctor, ought to be given it. This, in a nutshell, is what some of the critics of EBM are complaining about (see, for example, Greenhalgh, Howick, and Maskrey 2014; Kristiansen and Mooney 2004; Rasmussen 2004).

Some have rallied to the defense of EBM by pointing out that EBM has no intention of robbing clinicians of their role in helping themselves and their patients to work out whether a particular treatment is truly suitable, as a patient may have her own priorities and her own values. For instance, Akobeng (2005) points out that in weighing up the potential benefits against the potential harm of the "best" available treatment, different patients and carers may decide differently depending on their own personal/social beliefs and values. In some cases, their doctors might not feel it worth their while to raise the matter of systematic reviews of a particular treatment, if that treatment is not funded by the hospital/practice to which the patient belongs, or when it is obvious that their patients are not insured/too poor to opt for treatment under private health care.

Although this is valid, it does not, however, exhaust the import of the criticisms leveled at EBM. The crux of the matter lies elsewhere: one ought to distinguish between two different issues or criteria of relevance that have been conflated, as already shown, namely, clinical relevance and statistical relevance. The latter is necessarily expressed in statistical terms; the former has to do with specificities of the patient's medical biography. Clinical relevance may be helped by taking into account statistical relevance; however, as the two criteria are not identical, statistical relevance may in some cases be only part of the story for clinical relevance, and in other contexts, may even be of no help, as already observed earlier.

Increasingly, it appears that the criterion of statistical relevance is being better served by RCTs than that of clinical relevance, as statistical findings lend themselves readily as tools for the financial management of hospitals, for policy makers who have to allocate limited resources to run medical establishments and so on. The primary concern of these organizations is not about the care of the individual patient with his or her own specificities and medical requirements but with the care of patients as general groups or categories. That is why the difference between the two contexts may be put as follows: statistical relevance is concerned with a general disease called X, and with a general treatment for the said general disease considered to be the "best" available as determined by RCTs and EBM, while clinical relevance seeks an answer to the quite different question about what is the "best" for this particular patient, a treatment that could fall outside of the "best" as determined by RCTs and EBM.

The flaw complained about from the standpoint of the field clinician lies deep in the design of RCTs themselves (and, therefore, by implication of EBM as well). I have argued that their basic axiom is that of *homogeneity*. Only such an axiom can secure that reliable statistical findings be obtained; such findings are pertinent and relevant to decision-making on the macroscale of patient care, though not on the microlevel of patient care—ex hypothesi, *homogeneity* excludes specificities and particularities of individual patients.

RCTs AND THEIR LINEAR, MONO-FACTORIAL MODEL OF CAUSALITY

The *axiom of homogeneity* and the linear concept of causality go hand in hand as the monogenic conception of disease (see Lee 2012) upholds one disease/one cause, one effect. The linear model says the causal arrow goes in one direction only, from (one) cause to (one) effect (which contrasts with the nonlinear, multifactorial notion of causality, propounding a reciprocal relationship between cause and effect, with negative and positive feedback loops as well as synergistic effects). The billiard-ball model of causation embodies the linear model—the billiard cue (a bit of matter operated by the agent, the billiard player) imparts motion to the first targeted billiard ball on the table (another bit of matter), which, in turn, imparts motion to the second billiard ball (yet another bit of matter) upon hitting it, and so on. (For an excellent diagrammatic representation of the differences between linearity and nonlinearity in one example in the history of scientific explanations, see Bechtel and Richardson 1992.)

In an analogous manner, the disease entity, like the billiard ball, is a bit of matter—be it bacterium, virus, fungus, parasite, (faulty) DNA sequence—acting as cause, producing an identifiable effect. For example, some bacteria (the cause) act, say, on the guts (a bit of matter), imparting motion to other bits of matter in the guts, leading hence to the effect, namely, diarrhea. This is what is meant by saying that the causal model embodied in the dominant conception of disease in biomedicine is linear and mono-factorial. Basically only two things are involved—X and Y, X being identified as the cause and Y the effect. Such a relationship may be represented simplistically as: X→Y. This is summed up by the dictum "one bacterium, one disease" (the tubercle bacillus [*Mycobacterium tuberculosis*]/tuberculosis), "one virus, one disease" (the Ebola virus/Ebola virus disease, also called Ebola hemorrhagic fever or more simply Ebola), "one faulty gene, one disease" (the extra 47 chromosome/one set of developmental effects associated with the Down Syndrome [Trisomy 21]—see Lee 2012, chapter 9).

However, this model of causality may not work at all except via a definitional fiat (Bhopal 2008, 103–104). To make the point clearer, take peptic ulcer,[2] in which the bacterium *Helicobacter pylori* is singled out as the cause and peptic ulcer as the effect. Empirical evidence shows that *H. pylori* is not a sufficient condition of peptic ulcer; to say that it is entails that every person who has the bacterium in their stomach (as ascertained by endoscopy) would have the disease called peptic ulcer. World medical statistics show that 90 percent of people known to have *H. pylori* do not develop gastritis or peptic ulcers; only 10 percent do in the presence of the bacterium while 80 percent are asymptomatic. Though not a sufficient condition, is *H. pylori* a necessary condition? It seems not either, as 10 percent of peptic ulcer sufferers in the world suffer from it not because of *H. pylori* but because of their use of NSAIDs (nonsteroidal anti-inflammatory drugs such as aspirin, ibuprofen, naproxen sodium)—30 percent of regular NSAID users have one or more ulcers. This holds true in the developed economies in the world. It is true that 90 percent of peptic ulcers in the world are caused by *H. pylori* infection—this happens, by and large, in the developing economies (see Lee 2012).

The *H. pylori*, example, in particular, raises a lot of disquieting issues: How could a factor X that fails to be either a necessary or a sufficient condition of Y in all cases be said to be the cause of Y? It shows that the model of causality invoked to explain these phenomena is flawed and yet it is this model that is invoked by RCTs (and therefore also EBM). It can only be said to be "the cause" through definitional fiat (see also Lee 2012, chapter 9).

THE PLACEBO EFFECT AND DUALISM

Philosophically speaking, biomedicine[3] may be said to rest, at least, on two main philosophical/methodological struts: (1) the linear, mono-factorial model of causality looked at above and (2) dualism.

These two struts could (simplistically) be said to have been laid down by two great intellectual giants in modern European thought. The first strut is part and parcel of the new worldview, which came into existence with the birth of Newtonian physics (Newton's three laws of motion) and hence of all Newtonian sciences, including modern medicine. Newtonian science investigates macrosized objects within three-dimensional space. The human body, which is the object of study of modern medicine, is a macrosized object whose behavior in the main is understood within the mono-factorial, linear model of causality.

However, preceding Newton, was another intellectual giant of modernity, Descartes, who struggled to reconcile two things that at first sight seemed irreconcilable. These were the uniqueness of human beings, on the one hand, and their commonality with the (higher) animals, on the other. Humans, like animals, have bodies but they also have minds (variously called reason or soul), which distinguish them from animals. Dualism permitted Descartes to cope with this conundrum. All bodies are forms of matter; matter is brute, subject only to the laws of motion understood by the new (Newtonian) physics. Under the old dispensation, reason/soul resided in matter. Hence, eliminating them makes matter *par excellence* the fit object of scientific study. That is why for Descartes, animals are mere automata as they are only bodies with no soul.[4] For Descartes, then, humans are unique in two related ways: to possess not only body but also soul/reason/mind; mind and body, nevertheless, are two very different substances, with mind being superior to body. In this way, Descartes succeeded not only in rendering human bodies as matter-appropriate objects of scientific investigation, but also in satisfying the theological requirement that human beings uniquely have souls that have to be saved. As souls have nothing to do matter, they are beyond the domain of science and its methodology. In this way, Descartes paved the way for science and theology to coexist, even if not always, peacefully. Cartesian dualism permits interaction between mind and body, a view that modern medicine rejects in favor of epiphenomenalism, whose details will be explored in some detail a little later.

Dualism entails a hierarchy of status between the two halves of a dualistic pairing—in the pairing, humans/nonhumans, man/woman, humans are systematically privileged over nonhumans, men over women. This, Plumwood says, involves "hyperseparation" or "radical exclusion":

For distinctness, for non-identity or otherness, there need be only a single characteristic which is different, possessed by the one but not the other, in order to guarantee distinctions according to the usual treatment of identity .... Where items are constructed ... according to dualistic relationships, however, the master tries to magnify ... the number and importance of differences and to eliminate or treat as inessential shared qualities, and hence to achieve a maximum separation. ... denial or minimisation of continuity is important in eliminating identification and sympathy between members of the dominating class and the dominated, and in eliminating possible confusion between powerful and powerless .... A major aim of dualistic construction is polarisation, to maximise distance or separation between the dualised spheres and to prevent their being seen as continuous or contiguous. ... A further important feature of dualistically construed opposition is that the underside of a dualistically conceived pair is defined in relation to the upperside as a lack, a negativity. (1993, 49–52)

In the mind/body dualism, mind would then be superior to body, as mind is said to possess reason/soul whereas body is mere matter without any hint of reason to it. However, it soon became obvious that this version of dualism does not quite fit the agenda of modern medicine/biomedicine, although it did permit space for the scientific study of the body without theological interference. The version of dualism that would be helpful is ironically that which makes body superior to mind. It looks as if that the general Cartesian doctrine of dualism can logically support two contradictory positions. However, in reality, there is no contradiction once it is realized that the fundamental claim made by the Cartesian doctrine is simply that one party of the dualism (it does not matter which) is privileged or superior to the other party named in the dualism. Modern medicine/biomedicine, under hyperseparation, privileges body over mind—body is matter, but mind is not and matter takes precedence over non-matter. It is this version that then underpins the medicine, which, in turn, entails reductionism that mind must be reducible to matter, the mental to the physical (to brain and its activity).

Brain activity could be seen to be at work in an important domain of biomedicine, in psychopharmacology. When mental disturbance in an individual leads to socially undesirable/difficult behavior, the preferred strategy for intervention is not *via* psychotherapies of one kind or another (such as psychoanalysis, counseling, and so on) but through physical interventions, such as electro convulsive therapy (ECT) or psychopharmacologic drugs. Such drugs and ECT are deemed to be the way forward because the chain of causation runs straightforwardly from matter (the chemicals in the drug or the electric forces in ECT) to matter (other chemical/electric reactions in the brain), in accordance with the billiard-ball model.

In other words, pharmacology in biomedicine operates within a causal framework with the following characteristics, which permit some room to accommodate psychopharmacology as a subcategory of general pharmacology:

1. The causal arrow runs from matter (as drug) to matter (body)—if the patient suffers from inflamed gums, an antibiotic would eliminate the bacteria-based infection.
2. The causal arrow may run from matter (as drug) to produce effects at the level of mind. This is how psychopharmacological drugs work. A patient who suffers from depression (a mental phenomenon) would be prescribed an antidepressant (a bit of matter), which is expected to lift the depression by having an effect on other bits of matter in the brain. This model of states of consciousness (mind) appears to accord with what is called epiphenomenalism as opposed to interactionism (though not necessarily of the Cartesian variety at the level of different substances). The main difference lies in this. Under interactionism, mental states can bring about changes in physical states just as changes in physical states can produce changes in mental states—for instance, I decide to punch you in the nose because you have just insulted me (a mental matter), the punch goes out and as a result, your nose is broken and is bleeding badly (physical matter); I take a psychoactive drug (a physical matter) for my depression; my depression is then lifted (mental matter). Under epiphenomenalism, however, the causal direction is only from physical states to mental states, but not vice versa. Take this example: A mother is worried that her son may get killed fighting at the front (mental state), her blood pressure goes up (physical matter). News then arrive to say that her son is on his way home as the war is now officially over; she feels greatly relieved (mental state), her blood pressure returns to normal (physical matter). Such outcomes are not compatible, strictly speaking, with epiphenomenalism. On the other hand, if an antidepressant is swallowed by the patient who then no longer feels depressed (a mental state), this is permitted and considered unproblematic by epiphenomenalism as it holds that while mental events are caused by physical events in the brain, they themselves have no effects on any physical events. This is why biomedicine has chosen to go down the route of psychopharmacology and is dismissive of other forms of interventions that appear to adhere to interactionism, at the level of the mental interacting with the material to produce physical effects.[5]
3. As the causal arrow does not and cannot run from mind to matter (body) as understood in biomedicine, this would explain the attitude of biomedicine to the placebo effect, which appears to stick like a bone in its throat. Ever since the end of the Second World War, the placebo effect has attracted both research attention and, in turn, skeptical attention. Critics, initially, doubted that the phenomenon was a genuine one and with justification

as it was true that the results of these early trials appeared not to have conformed to the strict methodological requirements of RCTs; later, these defects were remedied. However, skeptical criticisms still prevail even to this day (see, for instance, Hróbjartsson and Gøtzsche 2001 and 2010), although there is increasing recognition that the phenomenon is in reality quite widespread, even beyond the testing of medical interventions (Brody 1980; Shapiro and Shapiro 1997; Guess, Kleinman, Kusek 2000; Moerman 2002, 2006; Evans 2004; Kradin 2008; Benedetti 2014). Today, so much evidence has been collected that denying the existence of the placebo effect at the empirical level becomes itself unscientific; yet, it seems to run counter to epiphenomenalism as it appears to claim that mental phenomena can cause physical effects. Hence, biomedicine is left maintaining the Maginot Line that the causal arrow *cannot run* from mind to body. What is the force of this "cannot"?

The force of this "cannot" appears to lie in those who subconsciously adhere to materialism-cum-reductionism,[6] even in the face of increasingly strong, sustainable, and large amount of empirical evidence, which is methodologically above board but which appears to undermine them. Their exponents may also (subconsciously) realize that letting go of them would not be easy as RCTs and EBM depend on the ability to rule out the placebo effect in effective and efficacious medical interventions. Pharmacology in biomedicine rests on the premise that only drugs containing the active ingredient/chemical of say, a certain plant, could be said to be effective and efficacious—in other words, the matter in the drug is not inert, whereas ex hypothesi, the matter in a placebo intervention (such as a fake pill) is inert. As one can immediately see, to consider that something inert could produce an effective and efficacious response would be hard to reconcile with the specific materialistic framework within which biomedicine operates. This framework does not simply say that matter acts on matter, but more specifically that only a certain kind of matter (a bit of matter that contains a certain identifiable and identified chemical) could act upon the disease entity (another bit of matter containing certain other chemical(s)). In other words, matter lacking the appropriate chemicals, such as inert matter (plain flour) would not, and could not, act upon the disease entity to eliminate it. To admit that inert matter can achieve such a result would be as good as giving up the specific materialistic framework of biomedicine, the notions of RCTs and EBM, the twin gold standards that operate to guarantee the "scientificity" of the entire framework. (See Lee 2012, chapter 8, for a more detailed account of pharmacology as well as the placebo effect.)

However, "scientificity" is not entirely exhausted by adhering to these twin gold standards, as there is another wider aspect of "scientificity" that could perhaps be ignored in the short run, though not in the long run. And this is that modern science is held ultimately to be grounded in and compatible with empirical evidence. Admittedly, there is no firm logical line that can be drawn between sturdy, unwavering commitment to a theory/paradigm (see the Popper-Kuhn-Lakatos debate in Lakatos and Musgrave 1970) and the point of its collapse. All the same, counter-evidence and so-called anomalies can gather force over time to challenge the status quo. The placebo phenomenon may turn out to play this subversive role.

Take the so-called Copernican Revolution, which is often regarded as a defining moment in the emergence of modern science. Geocentrism was not without merit; in spite of the so-called ad hoc hypotheses (such as epicycles and epicycles upon epicycles) invoked to prop up the main hypothesis, it performed well enough for the purpose of astronomical calculations. Nevertheless, heliocentrism eventually triumphed and brought new advances in scientific understanding. Analogously, the results of the systematic study of the placebo phenomenon (not only in medicine but also in other domains) may play a subversive role of undermining in the long run the intellectual edifice of biomedicine resting on reductionism, the privileging of the physical over the mental, the twin gold standards as embodied in RCTs and EBM today. Their adherents by elevating the latter as gold standards and adopting an ostrich-like attitude to the existence of the placebo phenomenon, ironically, are only drawing attention to the phenomenon as a glaring anomaly to the biomedical worldview, which is wedded to privileging the material/physical over the mental aspects of the person. In this sense, the twin gold standards by being held sacrosanct in the temple of "scientificity" may eventually, ironically, be seen to be the "Trojan Horse," which serves to undermine the very temple they are expected to uphold.

CONCLUSION

There appears to be some challenges to biomedicine as it exists today with its criteria of "scientificity" as embodied in RCTs and EBM.

1. RCTs and EBM appear to suffer from weaknesses from the standpoint of their "external validity" as well as "internal validity." In the opinion of some of its major critics, the central flaw in external validity may be overcome with more determined political will to ensure that tests be conducted in accordance with sound procedure to prevent concealment of negative results, the source of funding, and so on, which could produce bias and misinformation. However, in the opinion of this contributor, "internal validity" and "external validity" are intimately interlinked; hence, the

distinction is not helpful. Both kinds of flaws are interrelated flaws and together they undermine scientific validity.

2. As EBM rests on RCTs, flaws in the latter affect the former.

3. RCTs blur the distinction between statistical relevance and clinical relevance, the conclusions for the former are necessarily couched in statistical terms making them more relevant to policy makers, funding bodies at the level of governmental agencies, hospital managers while failing to address the concerns of the end user, namely, the patients and their doctors looking after them who have their own respective specificities which are medical, psychological, social, and economic. A drug effective at the macrolevel for the idealized abstract patient may not be efficacious for a specific individual. In that sense, RCTs may not have all that much to offer to the field clinician.

4. RCTs have also been identified to suffer from two related flaws (implicating EBM): its reliance on (1) the linear, mono-factorial model of causality, which may also be called the billiard-ball model, upholding "one cause, one effect," and (b) the *axiom of homogeneity*, which ensures that the linear, mono-factorial model of causality would function but by masking the mismatch between the model and empirical findings. Diseases appear not to conform to (1), and these constitute anomalies for the model, which are then ignored; biomedicine ignores them by turning (1) into a tautological truth.

5. RCTs and EBM, on another empirical front, also ignore the placebo phenomenon, as taking it seriously may well undermine their embodiment of "scientificity," which dictates that in the body/mind dualist pairing, the physical is privileged over the mental and the mental is reduced to the physical. The causal arrow only points from matter to matter or at best matter to mind (under epiphenomenalism), but never mind to matter. This ostrich-like attitude to the wide-ranging placebo phenomenon may ultimately in the end ironically subvert the twin gold standards themselves, to make way eventually for another conception of the patient as person in which the dualist pairing of body/mind would be superseded.

NOTES

1. Kalanthini, however, did not use the terms "statistical relevance" and "clinical relevance."

2. For detailed discussion of other examples and related issues, see Lee (2012).

3. Biomedicine used to be called modern medicine after humoral medicine, beginning with the ancient Greeks, was abandoned in favor of "solid" medicine.

4. This extreme view, however, was not accepted by all contemporaries or near-contemporaries of Descartes, although it enjoyed widespread support till quite recently, as advocates of animal welfare and rights remind us.

5. See Howard Robinson (2012) for an account of interactionism and epiphenomenalism and William Robinson (2015) of epiphenomenalism.

6. This kind of philosophical commitment looms large in philosophy of mind literature; the representatives at extreme ends of the spectrum are Churchland (1986) and Kim (2005). The alternative term to "materialism" is "physicalism."

REFERENCES

Akobeng, A. 2005. "Understanding Randomised Controlled Trials." *Archives of Disease in Childhood* 90:840–844.

Angell, Marcia. 2004. *The Truth About the Drug Companies: How They Deceive Us and What To Do About It.* New York: Random House.

Bechtel, William and Robert C. Richardson. 1992. "Emergent Phenomena and Complex Systems." In *Emergence or Reduction? Essays on the Prospects of Nonreductive Physicalism,* edited by Ansgar Beckermann, Hans Flohr, and Jaegwon Kim. Berlin and New York: Walter de Gruyter.

Benedetti, Fabrizio. 2014. *Placebo Effects: Understanding the Mechanisms in Health and Disease.* Second Edition. New York: Oxford University Press, 2014.

Bhopal, Raj. 2008. *Concepts of Epidemiology: An Integrated Introduction to the Ideas, Theories, Principles and Methods of Epidemiology.* Oxford: Oxford University Press.

Brody, Howard. 1980. *Placebo and the Philosophy of Medicine.* Chicago: University of Chicago Press.

Cartwright, Nancy. 2007. "Are RCTs the Gold Standard?" *BioSocieties* 2:11–20.

Chalmers, Iain. 2006. "From Optimism to Disillusion about Commitment to Transparency in the Medico-Industrial Complex." *Journal of the Royal Society of Medicine* 99: 337–341.

Churchland, Patricia C. 1986. *Neurophilosophy: Toward a Unified Science of the Mind-Brain.* Cambridge, MA: MIT Press.

Evans, Imogen, Hazel Thornton, Iain Chalmers, and Paul Glasziou. 2011. *Testing Treatments: Better Research For Better Healthcare.* Second Edition. London: Pinter & Martin Ltd., 2011.

Evans, Dylan. 2004. *Placebos.* Oxford: Oxford University.

Goldacre, Ben. 2012. *Bad Pharma.* London: Fourth Estate.

Greenhalgh, Trisha, Jeremy Howick, and Neal Maskrey. 2014. "Evidence Based Medicine: A Movement in Crisis?" *BMJ* 348. doi: 10.1136/bmj.g3725.

Guess, Harry, Linda Engel, Arthur Kleinman, and John Kusek. Editors. 2002. *The Science of the Placebo: Towards An Interdisciplinary Research Agenda.* London: BMJ Publishing Group.

Hróbjartsson, Asbjørn and Peter C. Gøtzsche. 2001. "Is the Placebo Powerless? An Analysis of Clinical Trials Comparing Placebo with No Treatment." *New England Journal of Medicine* 344(21): 1594–1602.

Hróbjartsson, Asbjørn and Peter C. Gøtzsche. 2010. "Placebo Interventions for All Clinical Conditions." *Cochrane Database Systematic Reviews* (20 January 2010) 106(1).

Kalanthini, Paul. 2016. *When Breathe Becomes Air.* London: Bodley Head.

Kaptchuk, Ted J. 2001. "The Double-Blind, Randomized, Placebo-Controlled Trial: Gold Standard or Golden Calf?" *Journal of Clinical Epidemiology* 54:541–549.

Kim, Jaegwon. 2005. *Physicalism, or Something Near Enough.* Princeton: Princeton University Press.

Kradin, Richard. 2008. *The Placebo Response and the Power of Unconscious Healing.* New York: Routledge.

Kristiansen, I. and G. Mooney, editors. 2004. *Evidence-based Medicine in its Place.* New York: Routledge.

Kristiansen, Ivar S. and Gavin Mooney. 2004. "Evidence-Based Medicine: Method, Collaboration, Movement or Crusade?" In Kristiansen and Mooney (2004).

Lakatos, Imre and Alan Musgrave. 1970. *Criticism and the Growth of Knowledge.* Cambridge: Cambridge University Press.

Lee, Keekok. 2012. *The Philosophical Foundations of Modern Medicine.* Basingstoke: Palgrave Macmillan.

Mill, John S. 1974. *A System of Logic.* Toronto and London: University of Toronto Press, Routledge and Kegan Paul.

Moerman, Daniel. 2002. *Medicine, Meaning and the 'Placebo Effect'.* Cambridge: Cambridge University Press.

Moore, Andrew. 2009. "What is an NNT?" URL = http://www.medicine.ox.ac.uk/bandolier/painres/ download/whatis/nnt.pdf, Accessed 17/08/2015.

Pedersen, Kjeld M. 2004. "Randomised Controlled Trials in Drug Policies: Can the Best Be the Enemy of the Good?" In Kristiansen and Mooney.

Plumwood, Val. 1993. *Feminism and the Mastery of Nature.* London: Routledge.

Rasmussen, Knut. 2004. "Evidence-Based Medicine And Clinical Practice." In Kristiansen and Mooney.

Robinson, Howard. 2012. "Dualism." *Stanford Encyclopedia of Philosophy.* Editor. Edward N. Zalta. URL = http://plato.stanford.edu/cgi-bin/encyclopedia/archinfo.cgi?entry=dualism Accessed 27/07/2015.

Robinson, William. 2015. "Epiphenomenalism." *Stanford Encyclopedia of Philosophy.* Editor. Edward N. Zalta. URL = http://plato.stanford.edu/entries/epiphenomenalism/ Accessed 2015.

Rumsay, Deborah J. "What a p-value Tells You about Statistical Data." URL = http://www.dummies.com/how-to/content/what-a-pvalue-tells-you-about-statistical-data.html. Accessed 2016.

Shapiro, Arthur K. and Elaine Shapiro. 1997. *The Powerful Placebo: From Ancient Priest to Modern Physician.* New York: WW Norton & Co.

Thelle, Dag S. 2004. "Randomised Clinical Trials in Assessing Inferences about Causality and Aetiology." In Kristiansen and Mooney.

Chapter 10

The Legitimacy of Preventive Medical Advice

Is Knowing Enough?

Delphine Olivier

Confronted with suffering, we all feel that the situation causing the suffering should be dealt with. Granted, the means can be discussed, and the consequences of some medical interventions can be terrible to such an extent that doctors are deterred from intervening. It is nonetheless true that the medical ambition to relieve suffering is not a matter for discussion. What happens then when the ambition is no longer to ameliorate suffering but to prevent future disease? In the absence of disease, what is the legitimacy of medical recommendations?

The immediate answer is that, in the absence of suffering, medical preventive decisions are grounded in some knowledge about future health and disease. This knowledge could explain why preventive recommendations target healthy individuals and populations, or why they may be undertaken even when they lessen the current well-being of an individual (for example, think of the dramatic case of prophylactic surgeries): if doctors know what is likely to happen in the future, it seems that they are entitled to provide recommendations aimed at escaping future pathology. To put it differently, it looks like preventive medicine is the place where the relation between knowledge and action is reversed: confronted with pain, curative medicine has too often to admit its knowledge is insufficient; when trying to prevent disease, medical knowledge claims to be ahead of individuals who are naively ignorant of, or even not caring about, the dangers they may face. The fact that preventive medical advice, by definition, precedes suffering implies that knowledge about future disease and health plays a key role in the justification of medical recommendations.

In this chapter, I want to question the connection between anticipative medical knowledge and preventive actions: What is the value of preventive knowledge when it comes to justifying preventive decision? The possibility of a discrepancy between accurate prediction and efficient prevention has been a recurrent matter for concern[1]—for example, making reliable predictions about diseases doctors have no idea how to cure immediately raises obvious ethical dilemmas. Instead, this chapter will focus on the case where both predictive knowledge and preventive action are available: if medicine *knows* how to prevent future disease, does that mean doctors *should* implement preventive action and advice? The first step of this chapter is the acknowledgment of a dissatisfaction in current preventive practice, which leads to a reconsideration of the question of the legitimacy of preventive advice (section 1). Given this initial conundrum, I argue that medical preventive knowledge should be submitted to a critical examination for preventive medicine to be legitimate. This critical examination involves a recognition of the limits of the scientific knowledge available, but also a more philosophical approach, for which I hope to provide some indications. Two important dimensions of the preventive knowledge will be explored: the centrality of the notion of health (section 2) and a specific temporality based on anticipation (section 3). The overall aim of this chapter is to show that these dimensions should be analyzed and taken into account in order to regain a legitimacy that is sometimes perceived as fragile by doctors dealing with prevention.

CONFLICTING VIEWS: A MALAISE
IN PREVENTIVE MEDICINE

The Crusade for Prevention

In the first part of twentieth century, prevention was often considered to have entered a new era, thanks to new scientific foundations that had arisen from the discoveries in the field of microbiology. There was then a strong enthusiasm for preventive medicine: the scientific avoidance of disease seemed to be imminent. Some doctors did not hesitate to say that the very goal of medicine would be revised by the preventive approach: treating disease could be considered as a transitional step before medicine could enable people to *avoid* disease. Among the numerous examples of such an optimistic description of preventive medicine, let us pick out a clear statement written by a British public health officer, George Newman.[2] In his book published in 1932 and portraying the glorious history of preventive medicine, Newman wrote: "The ideal of Medicine is the prevention of disease, and the necessity for curative treatment is a tacit admission of its failure" (Newman 1932, vii).

This kind of statement is not isolated in the history of medical prevention. At the end of the twentieth century, investigations into genetic susceptibilities paved the way for a new project of "predictive medicine." Medical science was thought to become capable of estimating the individual risk of developing a disease. Once again, many assertions were made about the evolution of medicine toward its genuine goal. For example, numerous such claims are to

be found in Jean Dausset's appraisal of predictive medicine: "Thus, historically, curative medicine has become preventive then predictive although the
reverse route is nowadays followed: prediction precedes prevention which
itself precedes the cure" (Dausset 1996, 29). The preventive or predictive
objective has been presented as the nobler one for medicine: in this view,
curing disease can only be the second-best choice.

Although doctors began to consider prevention a more valuable aim for
medicine than cure, they also complained about the difficulties they had to
face in order to make this golden age of medicine happen. Making individuals not only embrace the preventive attitude but also put it into practice was
not the smallest one. Confronted with what they perceived as the laziness and
foolishness of individuals, doctors sometimes gave the impression they had
to engage in a crusade for health. In the first issue of the journal *Preventive
Medicine*, in 1972, one could read the following statement: "The task facing those of us committed to prevention is an enormous one. We must work
against the profound human tendency—in ourselves and in others—to concentrate on the pressing problems of the moment to the exclusion of actions
to avert problems in the future" (Steinfeld 1972, 10). And again, this is not
an isolated declaration: numerous doctors or public health officers have complained about the difficulty of convincing people to make the right decisions
for their future health.

When they do not lament over individuals' insouciance, preventive actors
wish they were not always given the secondary role. Criticisms of the relega-
tion of preventive tasks to the lowest of political and medical priorities are
a recurrent feature of speeches dealing with prevention. A good example is
to be found in a wonder expressed by the former president of the American
College of Preventive Medicine, Halley S. Faust:[3] "What explains why, as a
matter of fact, in our Western Health care systems—at least in U.S. health
care—we concentrate so many more resources on alleviating illness than on
preventing it?" (Faust & Menzel 2011, 4–5). Convinced that prevention is
the right thing to do, the persistence of a disequilibrium between the funds
allocated to prevention and those granted to treatment can only be a major
concern for Faust and for numerous physicians involved in the promotion of
preventive medicine. Their complaints convey the impression that the respect-
able preventive ideal faces unfair resistances and difficulties—prevention: the
ugly duckling of medical science.

When Doctors are No Longer Convinced that Prevention is Better than Cure

The heralds of preventive medicine often refer to the commonsensical observation that one would rather avoid any kind of harm rather than suffer and
then go for treatment. However, it is far from obvious that the preventive

attitude is unanimously and wholeheartedly praised. The enthusiastic claims quoted in the previous paragraph should not be misleading: a more thorough examination of the debates troubling the medical world in recent decades shows that one can easily find doctors highly suspicious when it comes to prevention: an increasing malaise seems a much more accurate description of the field of medical prevention than an unstinting support.

What should be clear is that a large part of the criticisms addressed to preventive medicine comes from the medical community itself, even though they sometimes borrow their critical concepts from analysts who are not doctors themselves. The phrase "disease mongering" was first used in the late 1990s by the medical journalist Lynn Payer: she meant to denounce the different tactics employed to literally generate new diseases—such as saying that something is wrong with a normal function, making a biased use of statistics in order to magnify the benefits of a medication, or promoting miraculous technological solutions. Interestingly, this concept is now to be found in some first-ranking medical journals and was extensively discussed in a special issue of *PLoS* in 2006 (*PLoS*, 2006). In a similar manner the concept of "medicalization," which was developed by the sociologist Peter Conrad in the 1990s, is now endorsed by some medical discourses. The *British Medical Journal* section "Too Much Medicine?" is entirely devoted to the matter of undue medicalization. Another recently successful concept in the medical world is "overdiagnosis"—which is not to be mistaken for false positives. It is employed to talk about those individuals classified as no longer healthy even though they only have benign problems or low risks.[4] Annual international conferences devoted to the prevention of overdiagnosis have been launched in 2013.[5] Finally, there is the concept of *quaternary prevention*, accepted by the World Organization of Family Doctors (WONCA) dictionary in 2003, which provides the following definition: "Action taken to identify patient at risk of overmedicalisation, to protect him from new medical invasion, and to suggest to him interventions, which are ethically acceptable." The target of this kind of prevention is no longer the impending threat of the disease: what should be prevented is the damage caused by preventive medicine itself (Bentzen 2003). "Disease mongering," "medicalization," "overdiagnosis," "quaternary prevention": this array of concepts that have been used in medical journals over the last few decades makes it clear that there is no consensus on the preventive ideal as the ultimate medical goal.

The fact that top-ranking medical journals or well-known scientists actively take part in the denunciation of preventive medicine abuse should not be considered merely as a curiosity in the landscape of these criticisms. It reveals something of the nature of the problem. The example of David Sackett's attacks on prevention is worth considering. In an article boldly entitled

"The Arrogance of Preventive Medicine," Sackett denounces an "aggressively assertive," "presumptuous," and "overbearing" medicine (Sackett 2002). In this brief and virulent denunciation, it should be noted that David Sackett does not say that scientific medicine is not the right way to go; he does not call for less science (that is, for an alternative approach to medicine); instead he is yearning for a *more scientific* preventive medicine. Sackett is known for his contribution to the development of evidence-based medicine (EBM), and his approach to prevention is clear: given the nature of any preventive recommendation, which interferes with healthy people's lives, evidence is *more* important than in curative medicine. According to him, many medical "experts" believe that prevention can be beneficial to individuals, but in too many cases, this claim remains unproven. The concern here is that preventive medicine without a thorough evaluation of the available evidence is dogmatic and does not satisfy the scientific standards that should be met by medical research.

All in all, these suspicions insist that we should be cautious about the alleged commonsensical benefits of prevention. The legitimacy of medical intervention on healthy individuals is far less obvious than that of curative medicine: "A stitch in time may in some cases be unnecessary and even harmful" (Godlee 2005, 330).

Toward a Critical Understanding of Preventive Knowledge

How can we understand these conflicting views? How can prevention be considered both as the ultimate medical goal and as something we should worry about? Given the features of the malaise mapped in the preceding paragraphs, we can say that no legitimate preventive medical decisions can be made without a thorough examination of the knowledge involved. To what extent can medicine say something about future health and future disease? How reliable is this knowledge? Does it apply to individuals? What does remain outside of the predictive possibilities of medicine? Such are the questions that should be answered in order to ensure that doctors are morally entitled to formulate their preventive recommendations.

To a certain extent, these questions still belong to the area of medical research and can be tackled with scientific tools, even though a consensus among the medical community is difficult to reach. The number of cancers induced by X-ray is still an object of concern and debates, and the estimation of a certain number of prophylactic therapies or surgeries (and the subsequent pain, discomfort, changes in life habits) following false-positive results in early detection procedures is also worrying (Heath 2009; Löwy 2010). These worries are actually tackled by physicians anxious about the potential harm induced by preventive medicine and dubious about its legitimacy and by the involvement of some EBM centers in the fight against an iatrogenic preventive medicine.[6] Therefore, the fact that no medical intervention is neutral and that all might have detrimental consequences that should be measured is taken into account by medical research, which seems more and more prone to challenge the idea that prevention can do no harm.

However, I want to argue that scientific tools are not likely to be enough in order to answer these questions. What is also needed is a critical examination: for preventive actions to be legitimate, not only should the scope and the limits of the knowledge involved be clearly stated, but an investigation should be made of the core medical concepts involved and of the implications of the temporality of preventive medicine. There is room for a philosophical investigation of preventive knowledge and the following sections provide some clues for a critical unfolding of this knowledge.

UNFOLDING PREVENTIVE KNOWLEDGE: OLD AGE AND THE NOTION OF HEALTH

A Confusing Situation: Prevention and Old Age

Let us begin our investigation with a specific case: prevention applied to the elderly. In developed countries, where a great part of the population has long life spans, chronic disease has turned into the main threat to health. Coronary disease, diabetes, and chronic lung infections have replaced cholera, tuberculosis, and smallpox. Confronted with this situation, it has been argued that new ways of action should be considered and that prevention is our best shield against the diseases of our time. However, preventing disease in old age does not go without serious side effects. The dangers of preventive decisions with elderly patients have been regularly denounced by Iona Heath[7] in the *British Medical Journal*. One of her main points is that doctors should face the fact every individual will die: this means they should resist the temptation to prevent every cause of death and instead focus on preventing premature deaths. Or, to put it differently, in an aging population we should not be surprised to observe that noncommunicable diseases are the main cause of mortality. We should realize that the true meaning of prevention is the avoidance of disease leading to premature death. Once you have reached and crossed the average life span—in a country where the population is in good shape—prevention will not delay death for ever. Worse, we should realize that prevention in old age means nothing else than selecting a way of dying. Preventing heart failure in an old person does not mean she will not die; it means she will die from something else, possibly from a condition that she would have deemed worse than a heart attack (Heath 2010).

As simple and bold as it is, this argument has powerful implications. According to Heath, this *memento mori* should help us see that the denying of preventive measures in old age is not an unfair discrimination against old people. It should also make us realize that the will to prevent any cause of death, whatever the age of the individual, leads to a slow but unavoidable destruction of publicly funded healthcare system (because the expense will be infinite) and to an unfair allocation of expenditures (because it shifts the focus from sick to well, from poor to rich).

In addition to that, if we go back to our initial question—the justification of preventive decisions by the availability of preventive knowledge—the limits faced by the preventive ideal in the case of old age reveal that knowing is not enough. A doctor may know that a heart disease is likely to happen, he or she may know that some medication will reduce the risk, but that does not mean he or she is automatically entitled to formulate a preventive recommendation or prescription to the individual concerned. Prevention with old patients makes it clear that there exists at least one situation in which not questioning our faith in the benefice of prevention is misleading. Contrary to the fierce defenses of prevention mentioned in the first section, it is by no means obvious that a preventive decision should always follow medical knowledge about future disease and health. But this case reveals something more: what is striking is that the limitations are directly connected to the definition of prevention and to its central medical values: if preventing disease results in a long-term deterioration of health, then the core medical objective of restoring or preserving health has not been reached. Therefore, a reflection on these values should be included in the assessment of preventive decisions. This is not tantamount to saying that the boundaries of legitimate prevention are extrinsic to medicine: they are tightly linked to the definition of the medical art. To put it differently, it would be a mistake to disconnect the epistemological analysis of preventive medicine from a concern for the values involved—and save these more practical questions for a strictly ethical investigation. Such a disconnection conveys the idea that physicians do possess an efficient preventive knowledge and that we should be careful about the way they use it, confronting their recommendation with some transcendent ethical values (such as benevolence). Instead, what is needed is a joint assessment of the preventive knowledge and its ability to effectively benefit to individuals in the long run. The question stated by Iona Heath (is prevention in old age equivalent to a selection of death profile?) is not—at least not entirely—an ethical question. It is a medical question in its own right. The concern is that preventive medicine could become counterproductive by damaging health.

Health: Philosophical Investigations

When I began this chapter, I quoted those enthusiastic appraisals of prevention forecasting a redefinition of the objectives of medicine. The definition of the central aim of medicine was at stake in these claims, and this should be taken seriously. It seems that the lesson prevention in old age can teach us is that the linkage between preventive knowledge and preventive action has to take into account the goals of medicine. Providing a very old person with medication in order to prevent heart failure, without confronting—difficult as it may be—the question "What for?" might lead to painful and unwanted results. Granted, listing the core medical goals leads to debate. Relieving suffering is obviously one goal, preventing disease can surely be included, but as soon as we explore the possible extension of what prevention means, trouble arises. Should we include the promotion of health? Should medicine try to enhance humans? Any tentative definition of medical goals has also to confront some practices such as cosmetic surgeries or contraceptive prescriptions, which have nothing to do with relieving or preventing pain.[8] That a clear and analytical definition of the medical goals cannot be easily provided does not count as an argument against the necessity of including a reflection on these goals in order to articulate preventive knowledge and preventive recommendations. Indeed, the very *unfolding* of the questions is crucial—not the providing of an unlikely once-and-for-all answer.[9]

Following this path, another question soon arises. Indeed, whether or not we include the enhancement of health in the goals of preventive medicine, a minimal requirement for a preventive objective is the preservation of health. This also means that a reflection on the definition of health has to be included in any preventive decision. The definition of health has been a recurrent matter of concern for academics belonging to the field of philosophy of medicine.[10] Given those debates, how can any provider of preventive advice, medication, or intervention waste time with a question such as "What is health?" Should not this question be left to idle philosophers? On the contrary, I want to argue that omitting a reflection on health—perplexing as it may be—leads to a good number of contradictions. Let us have a look again at the problem of prevention and old age. The problem pointed by Iona Heath comes from a failure to take into account that the health of elderly people involves a proximity to death and is not reducible to an estimated low risk of heart failure. Prevention at any cost with elderly people is doomed to fail because it does not take into account that, in the old, health is, by definition, closer to death than mid-life health.

Unfolding the question of the definition of health is therefore as necessary as reflecting upon the aim of preventive actions. Though an attempt to solve this question could prove to be vain, it is worth attempting to explore further the connections between medical concerns and the questioning of health. One of the recurrent preoccupations when it comes to prevention is the fact that it could be indirectly damaging, through the creation of anxiety. Telling someone that he or she should undergo regular screening for a given disease is tantamount to creating an awareness of the eventuality of a disease he or she would not have given a thought about otherwise. What is important here is that this anxiety should not be considered as a mere drawback that would be worth paying in order to avoid future disease: it is itself an *actual reduction* of health or well-being. Generating anxiety about future disease can be strongly detrimental to present health, and this explains why "opportunistic prevention" (which happens when general practitioners give their patients preventive advice even though there is no connection between this advice—a preventive one—and the reason why these patients went for the doctors—their being ill) has recently been targeted as a dubious medical practice (Getz et al. 2003).

Interestingly enough, these concerns about anxiety, opportunistic prevention, and potential reduction of health can be highlighted by some philosophical analyses on the definition of health. In his essay on the normal and the pathological, Georges Canguilhem recognizes health as tightly linked to a confidence in one's body's ability to cope with its surroundings, to impose its own vital norms on the environment, to overcome illness (Canguilhem 1989). Having in mind this analysis of health could be a good reason for being cautious with any medical recommendations that foster a constant anxiety for one's health and ruin confidence in one's health. To provide a second example of a philosophical stance, Ivan Illich's criticisms of medical institutions are also connected to a definition of health, which involves an intimate dimension: health is something that has to be experienced. Hence, delegating the assessment of health to external devices or institutions, and attempting to reduce health to a set of scientific measurements, is also a negation of the very nature of health (Illich 1974).

This proximity between philosophical reflections and various expressions of a current malaise in preventive medicine is also visible in what has been labeled the "paradox of health," sometimes referred to in order to explain why a preventive orientation seems to lead to infinitely increasing expenditures. In 1988, an American psychiatrist Arthur Barsky described what he called the "paradox of health": in spite of the improvement of objective measures of health status in America, surveys revealed a decrease in the subjective feelings of healthiness (Barsky 1988). Since then, other studies have shown that the healthiest countries are also the countries where a larger number of people consider themselves in poor health (Heath 2005). In this context, prevention can only be more and more expensive, with limited results to be expected, and a potentially devastating effect on the sustainability of health systems. Given those elements, the medical literature reveals an awareness of the necessity to resist the urge to invest blindly more and more money in prevention. In this context, philosophical reflections could provide the necessary arguments to make the case against any systematic preventive recommendation. Understanding the "paradox of health" makes it necessary to engage with reflections upon the genuine nature of health, disease, and death. Illich's attempts to integrate the personal experience of suffering, healing, or death in the very definition of health (Illich 1974), which finds further development in contemporary reflections on the definition of health and our attempts to find technological solution for every medical problem (Le Blanc 2004), are only one example of the possibility for philosophical investigations to highlight some medical debates.

By no means do we want to say that philosophy will provide doctors with definitive answers when it comes to the legitimacy of their recommendations. But it is likely that an awareness of the insufficiency of a strictly biomedical definition of health will help them face the difficulties raised by some situations. The various situations and criticisms we have examined so far strongly suggest that expurgating the definition of health from its existential dimensions quickly leads to medical paradoxes. It can only be hoped that a reintegration of these dimensions in the medical debates can help reverse the situation.

THE TEMPORALITY OF PREVENTION

Tautological as it may look, it should be remembered that preventive medicine seeks to anticipate disease and therefore involves a specific temporality. This temporal dimension is no less crucial than the notion of health in our understanding of preventive recommendations, and this final section will try to justify this claim.

In order to analyze the temporality of prevention, it is useful to broaden our perspectives: considerations about how technologies and political or cultural values shape various temporalities are of valuable help. In an article published in 2009, Adams, Murphy, and Clarke described some of the features of the "anticipation regimes" that, according to them, are a distinctive feature of our societies. Their article underlines the growing importance of "speculative forecast," notably at the intersection of technoscience and life. Throughout their analysis, they also insist on the fact that anticipation is not limited to a certain knowledge about future events: it leads to an "affective state" and makes "hope and fear … important political vectors." Therefore, anticipation creates a connection between present and possible futures: "Through anticipation, the future arrives as already formed in the present, as if the emergency has already happened." This intertwining between knowledge and affects, between present and future, also shapes political landscape: the anticipation of future disasters, the management of hopes and fears are, according to the authors, a feature of the way we deal with our present problems (Adams, Murphy and Clarke 2009, 248–249).

It appears that most of the descriptions given in this article provide a useful framework for an analysis of prevention and help us to understand its specific temporal nature. Indeed, it should be noticed that "speculative forecasts" are also given more and more importance in preventive medicine. Not only do we expect prevention to erase the major causes of disease or accident, but we also expect medicine to make accurate predictions about future health. At the intersection of technological development and medical science, so-called "predictive medicine" is sometimes described as a new step in the medical anticipatory possibilities. Looking for biomarkers, the promise of that medical project is to provide individuals with a predictive knowledge about their future health (Auffray, Charron and Hood 2010). Granted, "preventive biomedicalization," as Clarke calls it, is only one facet of our current preventive medicine, which includes a large variety of knowledge and practices. Nevertheless, anticipation is crucial to understand preventive decisions, and the implications of this statement, both on an individual and on a collective level, have to be clarified.

From the perspective of individuals making decisions about preventive medicine, it is important to understand how knowledge of a future outcome interacts with present affects and present decisions. The word "preventive knowledge" can be misleading, by conveying a false impression that medical science is able to provide certainties about future health outcomes, where instead it should be noted that prediction for a given individual hardly ever reaches the level of certainty. Most of the time, individuals can only be provided with probabilities regarding their future health, the meaning of which is anything but clear (Schwartz 2010). Preventive knowledge is always expressed in terms of risk, probabilities, uncertainty; the process of decision will thus require a balance between different possibilities in the future and will include a gamble about the uncertain onset of the disease, or the outcomes of a given preventive measure. This decision involves complex trade-offs between present and expected future, and psychological mechanisms such as anticipated regrets. Indeed, psychologists—and neuroeconomists—interested in decision-making have recently focused on the way we weigh not only different possible outcomes but also the expected emotional states that would arise for any of these outcomes. This mechanism helps understand why what looks like two similar outcomes are not always deemed equivalent: the difference can be explained by the anticipation of increased regrets in one case

(Zeelenberg 1999). We do not need to go into any further detail here; both the analyses of the probabilistic nature of preventive knowledge and the models of decision-making can reach a certain level of complexity, but the lesson to be drawn from them is clear. Preventive decisions requires the ability to deal with uncertainties, with hopes, with fears, with expected regrets, with present and future. Because of the role of personal preference, it is impossible to think that there could be a direct translation of any preventive knowledge into a preventive decision: the acknowledgment of the specific temporality of preventive knowledge can only be one condition for enhancing individuals' ability to deal with their present and future health.

"Regimes of anticipation are distributed and extensive formations that interpellate, situate, attract and mobilize subjects individually *and* collectively" (Adams, Murphy and Clarke 2009, 249): if we now turn to the political dimension of some preventive decisions, the analysis of anticipation regimes also proves to be useful. It can be argued that recent trends in preventive medicine focusing on individuals and relying on technological solutions are closely linked to a certain vision of society, where individual responsibility is empowered over collective health systems (Sicard 2005). Under various and promising labels, such as "P4 medicine" (preventive, predictive, personalized, and participative) or "predictive medicine," some aspects of prevention interact with a political system in which health and "wellness" can be a consumer good. An awareness of the interrelation between some facets of preventive recommendations and a broader political anticipation regime might be something we should be aware of in order to articulate knowledge and recommendation for preventing disease. The link between the temporality of prevention and its political dimensions can also be explored in a different way. We mentioned earlier that the preventive knowledge is often probabilistic. The probabilistic nature of this knowledge is visible in some public health concepts[11] such as the "number needed to treat/to screen/to vaccinate." This ratio is an estimation of the number of individuals that must submit themselves to a medication, a health check, or a vaccine in order to prevent one individual to suffer from a given condition. Such a concept is a clear indication that preventive decisions do not automatically follow medical knowledge: no scientific knowledge will ever tell people whether a number needed to treat is to be considered acceptable or outrageous—this involves a political dimension that should not be veiled.

Whether it be at the individual or the population level, any preventive decision should take into account the temporality and uncertainty of the knowledge involved: a probabilistic, future-oriented knowledge, which should be balanced with political and individual evaluation of the anticipated trade-off offered by the preventive solution.

CONCLUSION: MEDICAL GOALS AND
THE DYNAMICS OF MEDICINE

Is predictive medical knowledge enough to legitimate preventive medical advice? This chapter has explored several reasons why the answer to this question should be negative. The main idea is that, to be legitimate, preventive medical decisions not only require preventive knowledge; we must also clearly assess the limits of this knowledge and confront some broader conceptual questions. Having a definition of health or an accurate model for individual decision-making may not be a prerequisite for the kind of critical examination I described, but the very awareness of the rich nature of health, the human experience of temporality, or some political questions is a fundamental step toward legitimacy in preventive recommendations. I do not contend that these conclusions are specific to the field of prevention. It may be the case that a critical examination of knowledge (scope, validity, meaning of the probabilities, etc.), which includes a reflection upon central medical values and goals (preservation of health in a way that is meaningful for individuals and collectives) is required for any medical action. If that is the case, preventive medicine can serve as a good case for reflection. Since the intuitive justification of relieving suffering is lacking, prevention forces us to think about the legitimacy of medical decisions and the possibility for medical knowledge to fulfill that role.

The requirements we have explored in this chapter have consequences for the way we consider philosophical reflections upon medicine. If we agree with the idea that medical knowledge should include a clear consideration of the initial values of medical enterprise, a neat distinction between epistemology and ethical reflections is no longer justified. Any analysis of medical knowledge, any epistemological apprehension of medicine should integrate a consideration of the values involved. This statement also has consequences for our representation of medical progress. Our analysis implies that a reflection on the goals should be a key element in the shaping of future trends in preventive medicine. This statement can interestingly be confronted with some considerations on the mechanisms of scientific expansion, described by the last century epistemologies. Indeed, during the twentieth century, philosophers and historians of science questioned the impulses of scientific and technological growth. Some seminal works argued that the effective development of science is independent from any representation of the ends—it is not teleological. Thomas Kuhn's description of what he called "normal science" is a striking description of a science that is pushed forward by an internal dynamics, by its present configuration rather than by an explicit and well-defined goal (Kuhn 1962). Hans Jonas's description of the technological development rests on a similar assertion: technology is to be feared for the very reason that, at some point, it acquires a tendency to follow an autonomous development instead of being used to face specific human needs (Jonas 1974). Even if medicine was not directly studied by these scholars, and does not fall strictly either in the category of science or in the category of technology, it looks like some of the criticisms we have quoted in the first section of this chapter fit into that framework. For instance, when Didier Sicard[12] denounces the increasing tendency to look for technological solutions for every human problem, he acknowledges that predictive medicine is engaged in a technological path whose internal dynamics leads to new technological researches. It could be the case that modern medicine tends to grow according to an internal, self-maintained, dynamics. Nevertheless, the analyses we provided in this chapter warn against the counterproductive consequences of excluding a reflection upon medical goals in medical decisions: a medical knowledge that does not enable people to enjoy better health ceases to be a truly medical knowledge.

The legitimacy of preventive decisions is not complete unless fighting against a mechanism of expansion disconnected from a clear statement of the goals: on that condition, preventive medicine could avoid the cynical comparison with *Dr Knock*[13]—who perfectly understood how to tell healthy individuals they are not yet ill.

NOTES

1. The Huntington disease (a neurodegenerative and nontreatable disease for which there exists a predictive DNA test) is often referred to as a paradigmatic and problematic example in which medical action is tragically lagging behind accurate predictions. See Gargiulo and Herson (2007).

2. George Newman (1870–1948) held the function of chief medical officer from 1919 to 1935.

3. Halley S. Faust was president of the ACPM from 2013 to 2015.

4. "Medicine's much hailed ability to help the sick is fast being challenged by its propensity to harm the healthy. ... Narrowly defined, overdiagnosis occurs when people without symptoms are diagnosed with a disease that ultimately will not cause them to experience symptoms or early death" (Moynihan et al. 2012).

5. See http://www.preventingoverdiagnosis.net.

6. The University of Oxford's Centre for Evidence Based Medicine of Oxford and Bond University's Centre for Research in Evidence-Based Practice are both partners of the Preventing Overdiagnosis conferences (http://www.preventingoverdiagnosis.net).

7. A British general practitioner, Iona Heath was, until 2012, president of the Royal College of General Practitioners.

8. For a discussion of the goals of medicine, see Christopher's Boorse chapter in Giroux (2016).

9. Section 3 will provide further indication about the political and individual preferences that take part in the decision-making process and are the reason why multiple answers can be expected to such questions.

10. Providing an overview of this debate is out of the scope of this chapter. We can only advise any reader interested in this question to look at a recent publication (Giroux 2016).

11. This chapter deals with "preventive medicine" in a sense that include some public health measures. Since preventive medicine is not a medical discipline with clear-cut boundaries, our criterion for talking about prevention is the following: any medical action that takes place before the onset of disease counts as preventive.

12. Former president of the French "Comité consultatif national d'éthique."

13. *Knock ou le Triomphe de la médecine* is a French comic play written in 1923 by Jules Romain, depicting the growing power of Dr Knock over a village where everyone was—and felt—perfectly healthy before he arrived. This play is frequently referred to by French criticisms of the abuses of preventive medicine.

REFERENCES

Adams, Vincanne, Michelle Murphy, and Adele E. Clarke. 2009. "Anticipation: Techno-science, Life, Affect, Temporality." *Subjectivity* 28: 246–265.

Auffray, Charles, Dominique Charron, and Leroy Hood. 2010. "Predictive, Preventive, Personalized and Participatory Medicine: Back to the Future." *Genome Medicine* 2: 57.

Barsky, Arthur J. 1988. "The Paradox of Health." *The New England Journal of Medicine* 318(7): 414–418.

Bentzen, Niels (ed.). 2003. *WONCA Dictionary of General/Family practice*, Wonca International Classification Committee.

Canguilhem, Georges. 1989. *On the Normal and the Pathological*, Zone Books, New York.

Dausset, Jean. 1996. "Predictive Medicine." *European Journal of Obstetrics & Gynecology and Reproductive Biology* 65: 29–32.

Faust, Halley S., and Paul T. Menzel Paul (eds.). 2012. *Prevention vs. Treatment: What's the Right Balance?* Oxford Scholarship Online.

Gargiulo, Marcela, and Ariane Herson. 2007. "Le risque de la prediction." *Contraste* 26: 221–231.

Getz, Linn, Johann A. Sigurdsson, and Irene Hetlevik. 2003. "Is Opportunistic Disease Prevention in the Consultation Ethically Justifiable?" *BMJ* 327: 498–500.

Getz, Linn. 2006. *Sustainable and Responsible Preventive Medicine,* PhD thesis, Norwegian University of Science and Technology, Trondheim.

Giroux Élodie (ed.). 2016. *Naturalism in the Philosophy of Health: Issues and Implications*. Springer: Switzerland.

Godlee, Fiona. 2005. "Preventive Medicine Makes Us Miserable." *BMJ* 330.

Heath, Iona. 2005. "Who Needs Healthcare—The Well or the Sick?" *BMJ* 330: 954–956.

Heath, Iona. 2009. "It is Not Wrong to Say No." *BMJ* 338: b2529.

Heath, Iona. 2010. "What do We Want to Die From?" *BMJ* 341: c3883.

Illich, Ivan. 1974. "L'expropriation de la santé." *Esprit* 436(6): 931–940.

Jonas, Hans. 1974. *Philosophical Essays: From Ancient Creed to Technological Man.* Prentice Hall, Englewood Cliffs.

Kuhn, Thomas. 1962. *The Structure of Scientific Revolutions.* University of Chicago Press, Chicago.

Le Blanc, Guillaume. 2004. "Le Conflit des médecines." *Esprit* 284: 71–86.

Löwy, Ilana. 2010. *Preventive Strikes: Women, Precancer, and Prophylactic Surgery.* The Johns Hopkins University Press, Baltimore.

Mangin, Dee, Kieran Sweney, and Iona Heath. 2007. "Preventive Healthcare in Elderly People Needs Rethinking." *British Medical Journal* 335: 285–287.

Moynihan, Ray, Jenny Doust, and David David. 2012. "Preventing Overdiagnosis: How to Stop Harming the Healthy." *BMJ* 344: e3502.

Newman, George (Sir). 1932. *The Rise of Preventive Medicine.* Oxford University Press, London.

PLoS Medicine. 2006. *Special Collection: "Disease mongering"* 3(4), 2006.

Sackett, David L. 2002. "The Arrogance of Preventive Medicine." *CMAJ* 167(4): 363–364.

Schwartz, Peter. 2009. "Disclosure and Rationality: Comparative Risk Information and Decision-Making about Prevention." *Theoretical Medicine and Bioethics* 30: 199–213.

Sicard, Didier. 2005. "Les Perspectives de la Médecine Préventive et Prédictive." *Revue française d'administration publique* 113(1): 121–125.

Steinfeld, Jesse L. 1972. "Preventive Medicine: The Long-Term Solution." *Preventive Medicine* 1: 10–11.

Zeelenberg, Marcel. 1999. "Anticipated Regret, Expected Feedback and Behavioral Decision Making" *Journal of Behavioral Decision Making* 12: 93–106.

Chapter 11

Translational Research and the Gap(s) between Science and Practice

Treating Causes or Symptoms?

Marianne Boenink

For almost 15 years now, "translational research," "translational science," and "translational medicine" have been buzzwords in biomedical research policy in many countries. The terms roughly indicate both a way of thinking and a set of activities, all of which are thought to foster a link between the knowledge produced in scientific (lab) research, on the one hand, and clinical practice or health care more generally, on the other. Visions of translational research[1] have been put center stage in biomedical research funding programs in many countries, most famously in the National Institute of Health Roadmap for Medical Research of 2003 (Zerhouni 2003). Since then, many other countries have followed suit. Partly as a result, "translation" is now everywhere: in the names of research projects and research centers, in training programs, and in the title of journals and of publications. Maienschein and colleagues in 2008 already observed, in a critical vein, that translational research had become an *imperative* for those working in biomedical research: it simply seems the right thing to do (Maienschein et al. 2008). More recently, Harrington and Hauskeller (2014) repeated this observation. They added, however, that whereas the translational imperative dominates public and institutional perceptions of biomedical research, it still seems peripheral to and may not have changed much in actual biomedical research practices.

If the term is ubiquitous in discourse about biomedical research and apparently has a strong normative force, what is the buzz all about, then? In common language "translation" is, of course, first and foremost about the activity needed to enable communication between people speaking different languages. As a metaphor, it therefore suggests the existence of two different domains or communities having trouble understanding each other. The use of the "translation" metaphor in the context of biomedicine thus first of all signals an underlying *problem diagnosis*: it implies a disconnection that needs to be overcome. And indeed, translational research is commonly promoted as an effort to bridge a gap between two domains, roughly construed as "science" (or the "lab bench") and "clinical practice" (or the "bedside"). However, even though there is a clear "mainstream" view of what translational research is about, a number of alternative interpretations have been brought to the fore. As several authors have shown, both the underlying problem diagnosis (the gap to be bridged) and the solution offered (translational research) are interpreted in many, sometimes radically diverging ways (Woolf 2008; van der Laan and Boenink 2015; Vignola-Gagné 2013). One might even say that "translational research" has become a contested concept.

In this chapter, I aim to provide an improved understanding and a critical assessment of the recent "translation buzz." In particular, I analyze which presuppositions are implied in different views of translational research, regarding both the character and the status of scientific knowledge, and the way this knowledge can and should connect with medical practice. I discuss these presuppositions against the background of historical and philosophical insights in the relation between medical knowledge and medical practice. Doing so helps us to realize that translational research actually is the most recent way to express a long-standing concern with the imperfect connection between biomedical knowledge and clinical practice.

First, I analyze the different views of translational research by asking for which problem "translation" is supposed to offer a cure, showing that actually numerous gaps have been identified. Next, I review the perceived causes of these gaps and the measures proposed to counteract these causes. A crucial controversy here is whether science itself is or is not to blame for the gap. Whereas most proponents of translational research do not find much fault with science, a minority does. These latter critics might find support in history and philosophy of medicine where, as I will show in the third section, several authors have argued that the gap between research and practice is very much a product of the emergence of modern biomedical science. If we take this diagnosis seriously, however, even the solutions proposed by those critical of current biomedical science seem to have important limitations, as I argue in the fourth section. Seen from a historical and philosophical perspective, many proposals to further translational research underestimate how deeply rooted the gap between scientific knowledge and clinical practice actually is. Translational research risks then, to use a medical metaphor, treating the symptoms, rather than the causes, of the gap between scientific knowledge and medical practice.

A note on methodology: the first two sections are based on empirical material collected by Anna Laura van der Laan and our joint analysis of that material, published in van der Laan and Boenink (2015). We searched PubMed for scientific publications using the terms "translational research/translational medicine/translational science," covering the period 1993–2011. For a more elaborate description and justification of the set of documents and the methods used, see van der Laan and Boenink (2015).

TRANSLATIONAL RESEARCH: TREATING WHICH PROBLEM?

Pleas for translational research in general refer to a "translational gap" that needs bridging. The diagnosis of this gap, sometimes also labeled as a "lag" or a "valley," is not exactly uniform. First of all, views on where to localize this gap vary. In addition, proponents disagree about the precise character of the gap and what is problematic about it. Finally, and partly related to these earlier differences, the causes identified to explain the gap vary. As a result, the treatment that translational research is supposed to provide can take very different forms. Let us first distinguish the different diagnoses of the problem.

When it comes to the localization of the gap, the metaphor provides again a helpful starting point. The existence of a gap implies that there are two sides, but different authors identify these sides in different ways. Analyzing scientific publications about translational research identified via PubMed, we found that one side of the gap is often the same, but that views on the other side of the gap vary. This one side is usually firmly located in preclinical biomedical research, whether basic biochemical research, animal experiments, or other scientific approaches and methods. This research is thought to produce knowledge that *may be* relevant for human health, but the research itself did not involve any human body parts yet. Identifications of the second side of the gap range from first-in-man studies, via development of medical technologies, improvement of clinical practice, and health benefits for the individual, to public health benefits. "The" translational gap thus can be localized between, for example, preclinical research and clinical studies, but also between preclinical research and benefits for the individual.

Depending on the localization of the second side, the gap is thought of as narrow (and hence relatively easy to bridge) or as a very broad one (which may be hard to cross). Each of the alternative "sides" subsequently mentioned here gradually enlarges the gap with preclinical research, because the latter ones, like Russian dolls, include the earlier ones. This has led several authors to distinguish "subgaps." Woolf (2008), for example, distinguishes a gap between scientific knowledge and the development of useful applications (T1) and a gap between availability of application and actual benefit (T2). Khoury and colleagues (2007) propose to distinguish four translational gaps: T1 (from lab science to clinical trial), T2 (from trial to technology development), T3 (from development to implementation), and T4 (from implementation to effectiveness). The point of distinguishing subgaps is, as these authors stress, that even when preclinical work leads to clinical tests with positive results, this does not automatically lead to new technologies, clinical use of those technologies, and the aimed-for health benefits. Still, these authors seem to agree with those focusing on the more widespread narrow view of the gap that the connection between preclinical and clinical research is crucial for any improvement of health care.

Localizing a gap, however, is not necessarily identical to diagnosing a problem. The suggestion is, of course, that the two sides (whatever they are) need to be in closer connection than they currently are. Why? Presumably because this would enhance the realization of some goals or values. These values are, interestingly, always related to the second side of the gap, regardless whether this side is identified as clinical research, technology development, clinical use, or health impact. An improved interface between basic science and clinical research, for example, would contribute to more clinically relevant knowledge or technologies. More interaction between clinical researchers and health care practitioners is thought to result in a more successful implementation of innovations and possibly in improved health outcomes. There is, then, an imbalance between the two sides implied in any translational gap. The first side is mainly thought of as *instrumental* to the values of the second side. This also explains why critics of translational research feel that basic research is not valued in its own right (see, for example, Maienschein et al. 2008).

This instrumental view of preclinical research in recent pleas for translational research tends to go along, however, with a very positive evaluation of what it delivers. One could even claim that the diagnosis that the aims of the second side could be further enhanced is based on the idea that preclinical research has produced many insights that could be relevant to health. The problem, according to this view of the gap, is that these knowledge gains are not reflected in clinical research, in innovative health care practices, and/or in ultimate health gains. The argument that the aims of the second side of the gap could be enhanced is ultimately based on a positive assessment of developments in preclinical research. *If* the insights from preclinical science would be taken up properly in clinical trials and clinical care, the assumption seems to be, then ultimately health would benefit.

The problem diagnosis driving most pleas for TR, then, is that the knowledge and tools produced in preclinical research are insufficiently harvested by clinical practice, meaning they have less impact on health than they could. A typical argument is "The ever-expanding pool of knowledge garnered from basic science has not efficiently translated into clinical products" (Albani et al. 2010, 642). This type of argumentation became particularly strong after the Human Genome Project was completed and other "omics" approaches in science started burgeoning. The director of the National Institutes of

Health (NIH), after the NIH published its Roadmap for Medical Research, made it very clear that the problem was not with the basic science: "It is the responsibility of those of us involved in today's biomedical research enterprise to translate the remarkable scientific innovations we are witnessing into health gains for the nation" (Zerhouni 2005, 1621). A note of frustration surfaces in particular in papers focusing on the broad gap between preclinical research and actual use or health benefit. A recurring trope is that it takes too long before discoveries in basic science lead to approval and use of novel drugs and other therapies. Papers often mention that it may take fifteen to twenty years before a discovery leads to new drugs or new forms of patient care, and claim that paying attention to the "translational process" could substantially shorten this time span (for example, Davidson 2011). The German minister of research after announcing a translational research initiative was quoted as saying: "We want to get new treatments to the patients as fast as possible" (Kupferschmidt 2011, E219). There is confidence in basic and preclinical science, then, but much less so in what comes next. The causes identified to explain the slow or even absent uptake of preclinical research are numerous and varied, as we will see in the next section. They may include lack of facilities and funding, lack of communication between different experts, bothersome regulation, but also publication policies, lack of up-to-date clinical guidelines, or lack of awareness of user practices.

In many cases this diagnosis builds on a rather linear view of the innovation process in the biomedical domain. Whatever the end point (clinical research, clinical practice, [public] health gains), the connection between preclinical research and that end point tends to be framed as a pipeline, in which ideally a continuous flow occurs from one side to the other. To be sure, some authors point out that the innovation process can (or even should) also include feedback loops. This happens when, for example, clinical experience is the starting point for novel hypotheses to be investigated in the lab (see, for example, Weinberg and Szilagyi 2010) or when future users are engaged in the design process of a new medical device (for example, Rosenblum and Alving 2011). In most publications, however, the assumption is that translation occurs when there is an undisturbed, unidirectional flow from preclinical research to clinical research, clinical practice, and health.

Precisely because the pipeline is not flowing as smoothly as it should, additional activity is needed to move the insights from basic science further along. The metaphors used to explain what translational research is about tend to stress "pulling" and "pushing." The National Cancer Institute, for example, in the 1990s established the Specialized Programs of Research Excellence (SPOREs) to "promote interdisciplinary research and to speed the bidirectional exchange between basic and clinical science *to move basic research findings* from the laboratory to applied settings involving patients and populations" (cited in National Cancer Advisory Board 2003, 3, my emphasis). Harrington and Hauskeller cite a 2014 UK National Institute for Health Research document asking for researchers with track records in "translating advances in basic biomedical research into clinical research, and *pulling through* basic biomedical research findings into benefits for patients" (2014, 193, my emphasis).

To sum up, many proponents of translational research start from the observation that preclinical research has produced lots of potentially valuable insights, which for some reason do not reach the subsequent parts of the biomedical innovation pipeline. Their problem diagnosis suggests nothing is wrong with the source of the flow, but that "obstructions," "bottlenecks," or "roadblocks" in the pipeline are hindering a smooth flow from one side to the other (however defined).

CAUSES OF THE TRANSLATIONAL GAP(S)

As the metaphors of the bottleneck and the roadblock suggest, many proponents of TR when identifying the causes of translational gaps point to the pipeline, rather than to the flowing substance and its source. They focus on the organizational, financial, and regulatory *context* in which scientific research and development take place, not on the scientific work itself. In general, authors identify multiple causes for the gap(s) they perceive. Which specific causes they refer to depends, of course, on their characterization of the gap. Those focusing on the narrow gap between basic and clinical research (T1) identify fewer and less varied causes than those who also include the subsequent gaps between research results and clinical impact (T2, T3, and T4). With the broadening of the diagnosed gap, ever more potential causes come into view. Let me first list the causes that are associated with gaps between preclinical and clinical research, and then discuss those associated with the gaps between research and impacts on and of clinical practice (see also van der Laan and Boenink 2015).

When publications on translational research localize the gap between preclinical and clinical research (as the majority does), one of the main recurring observations is that setting up and carrying out clinical trials is hard. This is first of all because such trials are expensive. The academic institutions doing preclinical work do not always have the budget to perform such trials, and before the advent of TR, dedicated funding for clinical trials was scarce (Woolf 2008). To facilitate translation, then, targeted funding programs have been formulated, which enable the testing of preclinical findings in a clinical setting. In addition, and often as part of the funding requirements, collaboration between academia and industry is stimulated. However, even if funding is not the problem, conducting such trials may prove challenging, among others because strict ethical regulations limit recruitment of research participants (Newby and Webb 2010). Thus, deregulation is proposed. Others point out that it may be hard to recruit participants anyway (West and McKeown 2006; Nasser, Grady and Balke 2011), and propose initiatives to create standing cohorts of people willing to contribute to research, from which specific trials could then recruit participants.

Whereas lack of funding and overregulation hamper the actual carrying out of plans for clinical research, many authors also suggest that a flow of good ideas for clinical trials is lacking. Some point out that there are insufficient researchers capable of picking up the ideas from basic science and transform them into clinically oriented research. This lack of clinician-scientists (Vignola-Gagné 2013) may be due to the institutional settings of both research and health care, which do not facilitate, let alone encourage, clinicians or scientists to combine clinical work with research. Combining the two is not only practically challenging; there is also a lack of rewards and career perspectives. Thus, setting up specific training programs, in combination with offering more rewards and career pathways, is put forward as an important way to stimulate translational work (Sartor 2003; Schnapp et al. 2009). In addition, some authors point out that the communication between lab and clinical researchers is wanting. There seems to be a lack of meeting places or communication platforms where the two groups can exchange relevant research questions, ideas about, and results from each other's work. As a result, novel journals are instituted to improve communication, and it is argued that research results from both sides (lab and clinic) should be made more widely available, also when they are negative. Negative results may, after all, spark novel ideas and/or prevent repetition of unpromising research directions.

The minority of publications on translational research that (also) focuses on the gap between clinical research and impact on health care practices and health outcomes perceives, of course, different, additional causes of the gap(s). Here, lacking awareness in care practices of up-to-date scientific insights is an issue (Fontanaroasa and DeAngelis 2002), as are institutional brakes on changing existing practices. To increase awareness, the importance of communication platforms is stressed once more. Providing open access to scientific journals is promoted as facilitating translation, since it at least enables all actors in health care to take notice of recent insights and emerging possibilities (Albani et al. 2010). Another way to translate scientific insights into clinical practice is to develop clinical practice guidelines (Davidson 2011). Such professional guidelines are an important vehicle to move forward novel insights and technologies to everyday practice. When it comes to implementation of such guidelines, reimbursement of the novel drugs and technologies is important too, and here insurance companies and government policy come into view (Thrall 2007).

The common denominator of all the causes listed here is, as I suggested above, that they are located in the *context* of R&D. The recurring assumption in the majority of publications seems to be that it is clear enough which type of work needs to be done to further translation. They claim that the type of research necessary to further translation is not new, but that there are many factors at play hindering the actors involved to carry out this type of research. However, not all authors promoting translational research are convinced that the causes of the gap(s) are by and large external to scientific practice. A minority, in particular those wondering why many innovative biomedical interventions ultimately fail to have a serious impact on individual and public health, seems to be less optimistic that science can do the job, if properly facilitated. These authors point out that too often, even after the full innovation pipeline has been successfully traversed, innovation does *not* achieve the expected impact. If so, there is good reason to wonder whether the activities performed were the right ones to begin with. Are current ideas about what constitutes good scientific research, or R&D more broadly, really sufficient? Maybe causes *internal* to science play a role in the emergence of the translational gap(s) as well.

In contrast with the dominant view of what causes the translational gap(s), then, these authors suggest we should look critically at the *first* side of the gap under consideration, regardless whether it is identified as preclinical, animal research, or even clinical research (in T2). Authors pursuing this line of thinking take issue with, among other things, current in vitro and animal models, because they appear to be insufficiently indicative of what happens in the human body (Cripe et al. 2005; Mayer 2002). Mouse models for Alzheimer's disease, for example, mimic only a limited part of the clinical and pathological manifestations associated with this disease. As a result, positive findings in such mice have limited predictive value for what a novel drug will do in humans. To address this cause of the translational gap, more complex models have to be developed, both in tissue culture and in animals (Mojica et al. 2006; Opal and Patrozou 2009). Other authors point out that the gold standard for effectiveness research in the biomedical domain, the randomized clinical trial (RCT), is severely limited too (Helmers et al. 2010; Weinberg and Szilagyi 2010). Because of the need for controlled situations, experimental settings are often hardly representative of the complexity and messiness of real life. If a new intervention shows promise in this controlled environment, it may still fail to do much in real-life situations. Or, to put it in methodological terms, in RCTs internal validity is often prioritized above external validity. As a result, outcomes in scientific settings may substantially diverge from outcomes in broader use practices. These authors stress that we need different types of scientific research to narrow the gap between RCTs and health care practice. Research should be more often conducted in "real life" settings (Weinberg and Szilagyi 2010). Also, outcome measures used in such research should be representative of what is most relevant to patients. This means, for example, that it is not sufficient to show that certain molecular parameters decrease; what matters is that subjective experiences improve. The broader plea for patient-reported outcome measures is, from this perspective, also a plea for improved translation.

Finally, a number of authors point out that current ways of doing R&D are too specialized and insufficiently connected. This is not only an issue of organization of science, although it is usually bound up with organizational differences. The problem is also in the limited operationalizations of "disease" and "health" in scientific disciplines. Different types of research, for example, in cancer genomics and cancer epidemiology produce very detailed knowledge, but this knowledge is insufficiently shared and integrated because scientists insufficiently realize they are looking at similar objects. This cause of the translational gap(s) could be overcome by pursuing more integration of different types of observations in databases, which can then be mined for relevant correlations and novel hypotheses (Califf and Berglund 2010; Crist et al. 2004; Zerhouni and Alving 2006).

To sum up, the majority of authors in our data set did not extensively reflect on the question how the current conceptualization and approach of knowledge in the biomedical sciences links up with the characteristics and needs of clinical practice. The authors who do reflect on this issue suggest that the causes of the hampering innovation process may be due as much to the limitations of current research methods. If this is true, even successfully removing all blocks and bottlenecks in the innovation pipeline would not suffice to make patients profit from scientific insights. The problem is also in science itself. Interestingly, this observation may find support in work done by historians and philosophers of medicine.

THE PROBLEMATIC RELATION BETWEEN SCIENTIFIC KNOWLEDGE AND MEDICAL PRACTICE

The diagnosis that the gap between preclinical science and clinical practice may actually be partly due to the characteristics of current scientific knowledge links up with an observation that has since long been put forward by historians and philosophers of medicine. Historian Charles Rosenberg, for example, has argued that such a gap actually was born with the emergence of modern medical science at the end of the eighteenth and during the nineteenth century (Rosenberg 1977). This birth changed not only the location and actors involved in medical knowledge production; the type of knowledge strived for changed as well. Even the object of knowledge was framed in a different way than before; a change that could be summarized by the cliché that modern biomedical science treats diseases, not persons. And this shift in object might explain why scientific knowledge does not sit easily with medical practice. Without pretending to give a full historical account, let me highlight here some of the shifts Rosenberg and others have reconstructed (Rosenberg 1977, 2002, 2003; Jewson 1979; Sturdy and Cooter 1998; Pickstone 2000).

At the end of the eighteenth century, medical care in Europe was mainly provided by family doctors in private practices, visiting patients in their homes. Patients were mainly from the elite, since others could not afford this type of care. Doctors had to vie for clientele, and in the end the patient's subjective assessment of the quality of care was crucial. This meant that the subjective needs of patients were leading for the physician's activities. Research was embedded in the consultation and geared toward exploring the requirements and peculiarities of the patient and his or her situation. In the consultation, any aspect of a patient's situation could be relevant. Patients were approached as integrated, indivisible wholes of bodies, minds, behavior, environment, and history. Patients were supposed to present unique patterns of events. Hence, treatments working in one patient need not be relevant to patients with similar complaints. As doctors had to rely on their senses, their focus was on phenomena that both they and their patients could readily access: perspiration, pulse, urine, and the like. Treatment also focused on what Rosenberg calls "intake and outgo" (2002, 488). Manipulating what went into and came out of the patient's body (for example, by changing diet or climate, by prescribing drugs leading to profuse sweat, or by bloodletting), they tried to induce changes in functioning that restored the apparently disturbed equilibrium between body and environment.

As both Rosenberg (2002) and Jewson (1979) argue, the "disease" concept did not play a large role in this way of practicing medicine. Since disease was the result of the complex interaction between a huge number of elements, diagnosis required attention to the peculiarities of a particular patient and his or her situation. The type of knowledge relevant here was largely skills-based: it started with reading and listening to a wide variety of sources, including the patient and his or her relatives, and combining the findings into conjectures about intake and outgo, which would then be guiding the trial of interventions aimed to elicit specific responses. The knowledge gained about one patient was not easy to transfer from one patient to another, nor could it be taught to medical students in separation from the patient. Medical knowledge was inclusive and context bound, and could be gained only by caring for individual patients.

All this gradually changed when nineteenth-century thinking and doing started to be organized around diseases, instead of patients. Of course, this was not something happening suddenly; many factors contributed to the transformation. To mention just a few: the opening of human bodies (first the dead, later also the living) led to anatomical procedures and tools and enabled correlating of complaints with internal lesions. The emergence of hospitals stimulated the comparison of patients to each other, both in terms of symptoms and with regard to the lesions discovered when opening their bodies. The subsequent development of instruments to measure physiological functioning of bodies added to the idea that symptoms were not constitutive of the disease, but a secondary sign. Disease, that is, got an existence and a life of its own. It became an entity that could be studied independent from its manifestation in individual patients. The discipline of pathology devoted itself to the study of "disease" as such, in its spatial and temporal development.

Thus occurred what Rosenberg (2002) has called the "specificity revolution": a shift in medical thinking toward the idea that diseases should be conceptualized as *specific entities* with their own characteristics, which manifest themselves more or less clearly in, but exist independently from, sick individuals. Diseases in this view, like plants or animals, can be categorized into distinct groups, *species*, with distinct characteristics. This means that "disease" can be studied separately from individual patients. Ironically (and somewhat confusingly), the knowledge produced by such research is *generic*, rather than specific. Since the specificity revolution, then, the activity of research has been more and more decoupled from patient care. The newly emerging group of medical researchers aimed for knowledge about specific diseases that was generally valid and objective, in contrast with the highly patient-specific, ungeneralizable knowledge sought for in the previous era of medicine.

The pursuit of this general knowledge about specific "diseases" thus created a gap between the knowledge produced in scientific research and the knowledge needed in clinical practice. Here, after all, doctors still had to deal with individual patients and their peculiar complaints and stories, not with pure exemplars of a disease entity. The major challenge for doctors became to discern which "disease" is at play in sick individuals. Many different tools were developed to bridge the gap between generic knowledge and individual case. Nosologies listed the symptoms and signs distinguishing one disease category from another. Documents were written that standardized disease nomenclature (like the International Classification of Diseases). The procedure of "differential diagnosis" and other algorithms were developed to guide doctors' reasoning. None of these instruments, however, succeeded in fully closing the gap between general knowledge and individual cases. The observation that doctors in practice need clinical judgment in addition to generally valid, objective knowledge, has not lost topicality (for an overview and assessment of recent discussions on this topic, see Solomon 2015).

It is important to note that the emergence of the gap between scientific knowledge production and medical practice, resulting from the gradual emergence of medical science, has not gone unnoticed. On the contrary, it has been extensively criticized ever since it emerged. Rosenberg reports how U.S. clinicians in the mid-nineteenth century greeted the novel insights by French and German scientists with skepticism. He cites New York physician Alexander Stevens, who in 1836 wrote about French medical scientists: "They lose more in Therapeutics than they gain by morbid anatomy—They are explaining how men die but not how to cure them" (cited in Rosenberg 1977, 502). In a similar vein, nineteenth-century physicians in the UK warned that logical explanations and statistics can never replace "the one touchstone of practical medicine, experience" (Pye Smith, cited in Lawrence 1985, 511). In the first half of the twentieth century, critics argued that medical knowledge focused too much on tissue structures and forgot to teach students about patients as personalities functioning in environments (Brown 1998). And to mention just one more example, in the 1970s the American psychiatrist George Engel developed his biopsychosocial model of disease as an explicit alternative to the "reductionist" focus on bodily functioning in biomedical science (Engel 1977).

I do not want to suggest that all these criticisms of the medical sciences, voiced in different eras and contexts, are essentially the same. Many different aspects and motives are at play. A recurring theme in many critiques, however, is that the medical sciences lack sufficient relevance for and have limited use in medical practice. Contemporary authors worrying about a gap between medical science and practice are thus standing in a long tradition of voices critical of medical science. The calls for translational research express in a novel way concerns that are as old as medical science in its modern form, and they shift these at the same time. To see what these shifts imply, we need to more explicitly compare the concerns in discourse about translation with the earlier concerns.

TRANSLATIONAL RESEARCH IN A HISTORICAL AND PHILOSOPHICAL PERSPECTIVE

We just saw that from a historical perspective, a gap between medical knowledge and clinical practice emerged when medical science rose as an activity separate from patient care. Historical and philosophical reconstructions also show that this gradual divergence of medical knowledge and medical practice was more than a separation of activities and roles. It was intertwined with a differentiation with regard to the *object of study* (from patient to disease) and the *type and function of the knowledge* about this object (from guiding pragmatic intervention in individual patients to general, etiological explanations of disease). How do this gap and the associated explanations relate to the gaps and presumed causes put forward by proponents of translational research?

The long-standing gap between medical knowledge and medical practice is in many ways reminiscent of the gap(s) referred to in recent discourse on translational research, but there are also differences. Narrow framings of the translational gap tend to focus on gaps within research. They are concerned about the transition from preclinical to clinical work, or from knowledge to technology development. Broader, more inclusive framings of the translational gap, in contrast, do question the link between medical science and medical practice. These broader framings fulfil an important role in reminding those concerned only with the narrow gaps that improvement of health care and health requires more than knowledge production and technological innovation. Seen against the historical background sketched above, however, these broader conceptions themselves may ultimately be too limited as well. At least three observations can be made.

First, the majority of authors worrying about a gap between preclinical research and clinical practice and impact (however delineated) approach this gap as a lack of uptake of biomedical knowledge in clinical practice. They do not question the relevance of the insights produced in biomedical science to clinical practice, and locate the causes of the lack of uptake in factors more or less external to scientific knowledge production. The main point of the historical reconstruction, however, is that knowledge of disease is inherently insufficient to guide medical practice. First of all because the *type of knowledge* produced in medical science does not fit the character and aims of the clinical consultation, and secondly because the *approach to disease* taken in medical science is seen as too limited to do justice to the phenomena central to medical practice. The assumption in most translational research discourse that insights from science are relevant and useful for practice may be way too optimistic.

Secondly, there is a minority of authors writing on translational discourse who do look critically at science itself. They consider seriously the possibility that the lack of uptake may be due to a lack of relevance. This subset interprets the translational gap as at least partly the result of science itself. This implies that we can bridge the gap if we transform medical science in such a way that it will be more relevant to medical practice. These authors criticize in particular the reductionism in current biomedical science. They point out, for example, that the step from animal studies to reproduction of findings in humans is far from self-evident. More, or different, intermediary steps seem to be necessary to account for the complexity of human bodies.

This approach of translational research links up with one strand of earlier criticism of the gap between science and practice: the one focusing on the limited, often reductionist approaches of disease in biomedical science. Whereas reductionism, all things considered, may be a very fruitful strategy in scientific knowledge production, this does not always lead to knowledge that is easily applicable in the messy reality of everyday lives. With the ongoing molecularization of biomedical research, reductionism has reached new heights, and integration of all these insights has become ever more challenging. Seen from this perspective, then, the gaps between clinical knowledge and clinical practice and impact, and even between preclinical scientific knowledge and clinical knowledge, are symptoms of a deeply entrenched, but sometimes unproductive tendency in biomedical science. And these gaps may never be truly bridged if medical science does not change track, for example, by taking more holistic, systems-oriented approaches to disease (and health).

However, and this is my third and last point, the brief excursion to the birth of modern medical science and earlier critiques of this development also shows that the gap between knowledge and practice to a certain extent is part and parcel of the emergence of biomedical science. It is *not only* due to the reductionism in science, which might in principle be repairable by taking more inclusive approaches to disease. A richer, inclusive concept of disease will do nothing to overcome the gap between the type of knowledge needed when caring for individuals and the knowledge of disease in general produced in science. The historical reconstruction of the gap between medical science and practice suggests that this gap also emerged because of *the focus on "disease" as a phenomenon independent from patients*. In addition to reframing our concept of disease to move away from reductionism, we might need to reframe our conception of disease in a direction that does more justice to the highly individual character of the phenomena at stake. Although the term "personalized medicine" (actually another buzzword in contemporary medical research) seems to promote related ideas, this is not the direction I am looking for. Most visions of personalized medicine boil down to more fine-grained stratification of patients and are still about producing knowledge claims with general validity (Boenink 2016). What I am thinking of is a reframing of disease and health as dynamic and ultimately normative, rather than descriptive phenomena, as for example developed by Canguilhem (1991). He points out that whether a certain bodily state is problematic for a particular individual is not determined by objective criteria; it depends on whether this state disturbs the balance between bodily capacities, behavior, and social and natural environment in such a way that the individual is not able to actively establish a new form of balance. What is an acceptable balance, moreover, is ultimately a normative issue, depending on one's view of the good life. Such a reconceptualization implies that knowledge about disease and health is highly particular and explicitly normative. It will also be hard to predict in advance of trying to what extent such knowledge can be transferred from one case to the next.

There is no space here to develop this in detail, but my main point is that a further reframing of disease in view of bringing it closer to care for individuals may lead us to ways of knowledge production that do not satisfy the dominant conception of science as pursuing generally valid and objective statements. Truly bridging the gap between medical science and practice, if possible at all, may require not only that we do science differently but also that we go beyond science.

CONCLUSION

The surge of attention to translational research in medicine shows and stimulates a renewed awareness that producing scientific knowledge does not automatically lead to improvement of care and to gains in individual and public health. As such, the efforts to improve the translational process are definitely laudable, since there are many ways in which attempts to improve care by doing research can fail. As I have shown in this chapter, however, such efforts often build on assumptions about knowledge and its relation to practice that may be too simple or outright naïve. In particular, the suggestion that preclinical science produces many valuable insights, which are not taken up in clinical research or in clinical practice due to organizational, financial, and/or regulatory shortcomings, is insufficiently critical of science itself. The historical detour taken here shows how the gap between medical science and practice is intertwined with the emergence of science as such and the concurrent ideas about the type of knowledge sought for and the type of object studied. Against this background, it seems highly unlikely that science and practice can be reconnected without changing the practice and position of medical science itself. If translational research is to offer more than a treatment of symptoms, we also have to reflect on the type of knowledge pursued in medical science and the framing of the object of that knowledge: disease and health.

NOTE

1. For ease of reading, from here on I will use the term "translational research" only. The analysis presented, however, also applies to "translational medicine" or "translational science" (when used in the biomedical setting).

REFERENCES

Albani, S., J. Colomb, and B. Prakken. 2010. "Translational Medicine 2.0: From Clinical Diagnosis-Based to Molecular-Targeted Therapies in the Era of Globalization." *Clinical Pharmacology & Therapeutics* 87(6):642–5.

Boenink, Marianne. 2016. "Disease in the Era of Genomic and Molecular Medicine." In *The Bloomsbury Companion to Contemporary Philosophy of Medicine*, edited by James Marcum [pages unknown yet]. London: Bloomsbury.

Brown, Theodore M. 1998. "George Canby Robinson and 'The Patient as a Person.'" In *Greater Than the Parts: Holism in Biomedicine 1920–1950*, edited by C. Lawrence and G. Weisz, 135–160. Oxford: Oxford University Press.

Canguilhem, Georges. 1991. *The Normal and the Pathological*. New York: Zone Books.

Califf, Robert M., and Lars Berglund. 2010. "Linking Scientific Discovery and Better Health for the Nation: The First Three Years of the NIH's Clinical and Translational Science Awards." *Academic Medicine* 85(3):457–62.

Cripe, Timothy P., Blythe Thomson, Thomas F. Boat, and David A. Williams. 2005. "Promoting Translational Research in Academic Health Centers: Navigating the "Roadmap"." *Academic Medicine* 80(11):1012–8.

Crist, Thomas B., Andrew I. Schafer, and Richard A. Walsh. 2004. "Translating Basic Discoveries into Better Health Care: The APM's Recommendations for Improving Translational Research." *American Journal of Medicine* 116(6):431–4.

Davidson, Andrew. 2011. "Translational Research: What does it Mean?" *Anesthesiology* 115(5):909–11.

Engel, George. 1977. "The Need for a New Medical Model: A Challenge for Biomedicine." *Science* 196(4286):129–36.

Fontanarosa, Phil B., and Catherine D. DeAngelis. 2002. "Basic Science and Translational Research in JAMA." *Journal of the American Medical Association* 287(13):1728.

Harrington, Jean, and Christine Hauskeller. 2014. "Translational Research: An Imperative Shaping the Spaces in Biomedicine." *Tecnoscienzia* 5(1):191–201.

Helmers, Sandra L., Victoria L. Phillips, and Gregory J. Esper. 2010. "Translational Medicine in Neurology: The Time is Right." *Archives of Neurology* 67(10):1263–6.

Jewson, N. D. 1979. "The Disappearance of the Sick Man from Medical Cosmology, 1770–1870." *Sociology* 10(2):225–44.

Khoury, Muin J., Marta Gwinn, Paula W. Yoon, Nicole Dowling, Cynthia A. Moore and Linda Bradley. 2007. "The Continuum of Translation Research in Genomic Medicine: How can We Accelerate the Appropriate Integration of Human Genome Discoveries into Health Care and Disease Prevention?" *Genetics in Medicine* 9(10): 665–74.

Kupferschmidt, Kai. 2011. "Germany Clambers Aboard Translational Research Bandwagon." *Canadian Medical Association Journal* 183(4):E219–20.

Lawrence, Christopher. 1985. "Incommunicable Knowledge: Science, Technology and the Clinical Art in Britain 1850–1914." *Journal of Contemporary History* 20(4):503–20.

Maienschein, Jane, Mary Sunderland, Rachel. A. Ankeny, and Jason. S. Robert. 2008. "The Ethos and Ethics of Translational Tesearch." *The American Journal of Bioethics* 8(3):43–51.

Mayer, Lloyd. 2002. "The Real Meaning of Translational Research." *Gastroenterology* 123(3):665.

Mojica, Wilfrido D., Ayesha Arshad, Sanjay Sharma, and Stephen P. Brooks. 2006. "Manual Exfoliation Plus Immunomagnetic Bead Separation as an Initial Step Toward Translational Research." *Archives of Pathology & Laboratory Medicine* 130(1):74–9.

Nasser, Nariman, Deborah Grady, and C.William Balke. 2011. "Commentary: Improving Participant Recruitment in Clinical and Translational Research." *Academic Medicine* 86(11):1334–5.

National Cancer Advisory Board. 2003. *Advancing Translational Cancer Research: A Vision of the Cancer Center and SPORE Programs of the Future.* Report of the P30/P50 ad hoc working group. Accessed June 23, 2016. http://deainfo.nci.nih.gov/advisory/ncab/workgroup/p30-p50/P30-P50final12feb03.pdf

Newby, David E., and David J. Webb. 2010. "Translational Research: A Priority for Health and Wealth." *Heart* 96(11):815–6.

Opal, Steven M., and Eleni Patrozou. 2009. "Translational Research in the Development of Novel Sepsis Therapeutics: Logical Deductive Reasoning or Mission Impossible?" *Critical Care Medicine* 37(1 Suppl):S10–5.

Pickstone, John V. 2000. *Ways of Knowing. A New History of Science, Technology and Medicine.* Manchester: Manchester University Press.

Rosenberg, Charles E. 1977. "The Therapeutic Revolution: Medicine, Meaning, and Social Change in Nineteenth-Century America." *Perspectives in Biology and Medicine* 20(4):485–506.

Rosenberg, Charles E. 2002. "The Tyranny of Diagnosis: Specific Entities and Individual Experience." *Milbank Quarterly* 80(2):23–60.

Rosenberg, Charles E. 2003. "What is Disease?: In Memory of Owsei Temkin." *Bulletin of the History of Medicine* 77(3):491–505.

Rosenblum, David, and Barbara Alving. 2011. "The Role of the Clinical and Translational Ccience Awards Program in Improving the Quality and Efficiency of Clinical Research." *Chest* 140(3):764–7.

Sartor, R. Balfour. 2003. "Translational Research: Bridging the Widening Gap between Basic and Clinical Research." *Gastroenterology* 124(5):1178.

Schnapp, Lynn M., Meridale Vaught, David R. Park, Gordon Rubenfeld, Richard B. Goodman, and Leonard D. Hudson. 2009. "Implementation and Impact of a Translational Research Training Program in Pulmonary and Critical Care Medicine." *Chest* 135(3):688–94.

Solomon, Miriam. 2015. *Making medical knowledge.* Oxford: Oxford University Press.

Sturdy, Steve, and Roger Cooter. 1998. "Science, Scientific Management, and the Transformation of Tedicine in Britain, c. 1870–1950." *History of Science* 36:421–66.

Thrall, James H. 2007. "Building Research Programs in Diagnostic Radiology. Part III. Clinical and Translational Research." *Radiology* 243(1):5–9.

Van der Laan, Anna Laura, and Marianne Boenink. 2015. "Beyond Bench and Bedside: Disentangling the Concept of Translational Research." *Health Care Analysis* 23(1):32–49.

Vignola-Gagné, Etienne. 2014. "Argumentative Practices in Science, Technology and Innovation Policy: The Case of Clinician-Scientists and Translational Research." *Science and Public Policy* 41(1):94–106.

Weinberg, Geoffrey A., and Peter G. Szilagyi. 2010. "Vaccine Epidemiology: Efficacy, Effectiveness, and the Translational Research Roadmap." *Journal of Infectious Diseases* 201(11):1607–10.

West, C., and S. McKeown. 2006. "Translational Research in Radiotherapy Trials." *British Journal of Radiology* 79(945):716–8.

Woolf, Steven H. 2008. "The Meaning of Translational Research and Why it Matters." *Journal of the American Medical Association* 299(2):211–3.

Zerhouni, Elias A. 2003. "The NIH Roadmap." *Science* 302(63–64):72.

Zerhouni, Elias A. 2005. "Translational and Clinical Science: Time for a New Vision." *New England Journal of Medicine* 353:1621–1623.

Zerhouni, Elias A., and Barbara Alving. 2006. "Clinical and Translational Science Awards: A Framework for a National Research Agenda." *Translational Research* 148(1):4–5.

Chapter 12

Enacting Adherence to HIV/AIDS Care

How Multiplicity in Medicine Becomes a Singular Story

Suze Berkhout

Medicine is not a coherent whole. It is not a unity. It is, rather, an amalgam of thoughts, a mixture of habits, an assemblage of techniques. Medicine is a heterogeneous coalition of ways of handling bodies, studying pictures, making numbers, conducting conversations. Wherever you look, in hospitals, in clinics, in laboratories, in general practitioners' offices—there is multiplicity.

Mol and Berg (1998)

You will kill somebody, if they don't take their meds, if you're, you know, if you're not monitoring them on a daily basis … if they miss. Some of them are on meds that are more susceptible than others. You can't miss two or three days, you just can't do that.

HIV Outreach Nurse

Adherence to medications has been described as one of the key reasons for suboptimal clinical outcomes in chronic health conditions, limiting the success of treatment for a wide range of illnesses in medical systems around the globe (WHO 2003). Defined broadly as the extent to which individuals endorse and follow through on medical advice, particularly that regarding the usage, dosing, and timing of prescription medications, adherence is a term that has largely replaced that of "compliance," for concerns that the latter denotes passivity as well as asymmetry within physician-patient relationships. Within the area of HIV/AIDS care, adherence to antiretroviral therapy (ART) has garnered extensive attention clinically and within medical research—adherence rates of greater than 90% are widely quoted in the literature as necessary to minimize the potential for developing drug-resistant strains of HIV and to diminish the likelihood that infected individuals pass on the virus through blood or sexual fluids (Osterberg and Blaschke 2005; Bangsberg et al. 2003).

But adherence within HIV/AIDS care is an interesting issue for reasons beyond its role in drug resistance and infectivity: adherence to ART is a domain where HIV risk categories taken to be constitutive of personal identity (e.g., female sex worker, "FSW," or injection drug user, "IDU") are routinely utilized within research literature and are employed in clinical settings as factors relating to adherence, and despite general consensus that adherence is important, it is also the case that, in practice, different clinicians and researchers envision adherence quite differently depending on their particular perspectives. As such, adherence is not *only* a significant concept in the field; it is a slippery concept insofar as it is imbued with meanings that extend beyond its strict definitions—meanings that relate to the identities of those enacting adherence in practice.

In this chapter, I trace notions of adherence as these operate within a group of street-involved individuals with substance use issues, living with HIV/AIDS. Given concerns about transmissibility of HIV infection among drug users, this is a group for whom a wealth of data and attention has been generated clinically and in biomedical research. I engage with this subject in order to take up the issue of multiplicity in medicine, arguing that different meanings are attached to adherence within HIV practices and that these meanings are mobilized in various ways within the spaces where clinical and research activities occur. A seemingly singular entity in medicine, writes Annemarie Mol, may in fact have numerous assemblages of methods and practices mobilized around it. This multiplicity reflects different ontolog*ies* of that supposedly singular phenomenon (Mol 2000; see also Law's discussion of Mol, in Law 2004, 45–59). I'll argue that in HIV/AIDS care, attending to practices reveals a much messier view of adherence than what follows from conventional discussions of it. Differences are smoothed over as the divergent meanings of adherence become coordinated into a body of coherent, conventional knowledge concerning drug users' adherence to ART that can be found in the clinical and research domains. The smoothness of the conventional model of adherence to ART belies the fragmented identities and fluid movements of the people and actions categorized as adherent versus nonadherent within the spaces where HIV care is enacted.

This erasure of multiplicity invites a further question: What has been left unsaid, unexamined, and unknown about adherence to ART? I go on to discuss contradictions and uncertainties that are typically left unexplored by the conventional framework of adherence within HIV/AIDS research and clinical practices. I argue that biomedical knowledge concerning adherence to ART relies on an emphasizing and, crucially, an *erasure* of certain actions, practices, and happenings. Thinking through this in relation to epistemologies of ignorance, I conclude that what is "known" about adherence to antiretroviral therapies is a complex, coordinated body of knowledge imbued with issues of epistemic authority and norms of credibility. Group identities and social imaginaries concerning who is adherent versus nonadherent to therapy shape this body of knowledge, ultimately limiting its scope and value.

THE FIELD SITE

In order to tell this story, I draw on empirical findings from 15 months of qualitative field research I conducted with women living with HIV/AIDS in inner city Vancouver, Canada. The women who participated in the study were frequently homeless; many used injection drugs (cocaine, heroin) as well as crack cocaine; they had varying levels of experience with ART, and each would have been categorized as nonadherent to their medications at various points in time (though adherent at other times). The women in the study were predominantly Indigenous, though the extent to which they identified as such varied. Other participants included (predominantly Caucasian) physicians, nurses, social workers, pharmacists, and epidemiologists, all working within the same inner city setting. The project was informed by practices common to ethnographic methodologies—I attended clinic appointments, emergency room visits, and hospital stays; followed outreach and street-based caregivers on their daily rounds; observed in hospital, clinic, and community health settings; and interviewed patients and caregivers and participated in conferences and workshops relevant to adherence in HIV/AIDS care. The data set also includes an extensive body of medical, nursing, and other literature relating to HIV/AIDS and adherence to HIV care.

This breadth of information, the details, the comings and goings, enable me to give an account of the different ways that adherence is performed in different settings of care within inner city Vancouver. Adherence, on my accounting, is a function of a multiplicity of materials and practices—daily nursing visits, frozen blood samples, pill boxes, blister packs, reports with red, yellow, and green codes telling which drug will work (and which will not)—all of which must be coordinated. Different adheren*ces* exist in different locations through these materials and practices, some of which are easily taken up and assimilated into a body of knowledge concerning HIV/AIDS, some of which are marginalized, sidelined, or erased altogether. Telling these stories reveals multiplicity, how it is managed, and how it is implicated in the production of knowledge in this area of medicine.

WHAT MEDICINE "KNOWS" ABOUT ADHERENCE

Field notes, October 14, 2008

I get a phone call from Missy, announcing she's in the hospital again, and I head up for a visit. She's been admitted for cellulitis and her doctors have re-started her antiretroviral medications. We're going to do another interview today. As we begin, she opens the drawer of her bedside tray table and rattles through a number of containers, pulling out two syringes filled with a thick yellow paste—so thick that it coats the sides of the container it's in. I find out later, from a ward nurse, that it's Nystatin, an oral thrush prophylaxis. It's meant to be squirted into the mouth using the pressure of the syringe plunger, so that none of the thick yellow goop gets wasted. If it were in a normal medication cup, it would just cling to the plastic. In the past, I've seen her nurses loading the paste into the syringe; it often prompts them to remark how nasty the stuff is, that they wouldn't want to take it either. Missy says it tastes bad, it looks like kid's paint and that's what it tastes like too. Chalky. She throws the tubes into the garbage, placing a paper towel that's already in the trash can over top of the pile, covering the syringes. She tells me, "I don't even know what it is." I tell her I think it's part of her prophylaxis meds. "I don't know. I hope not ... I guess I should be taking them," she says. "Instead of sticking it in the drawer?" I ask. She tells me that she has asked the nurses to make sure she takes it, because she won't otherwise.

In conventional accounts of health behavior, Missy's hiding her Nystatin in the garbage would be described as an act of nonadherence. Within the field of HIV/AIDS care, there is a strong interest in medication-related factors concerning adherence, in particular to ART. Typically, these factors include issues such as the size of pills, difficulty swallowing medications, number of pills, and challenges with timing of medicines (some requiring food, some an empty stomach, multiple times in a day). In Missy's case, the medication looks and tastes bad (not even the nurses would want to take it), suggesting that difficulties ingesting medicines are significant. When medications produce side effects (e.g., nausea, gastrointestinal upset, changes in body fat distribution) that require additional medications to control, the antiretroviral regimen quickly becomes onerous and difficult to establish as a daily practice (Stone et al. 2001; Bartlett et al. 2001; Conrad 1985). Medication issues are seen as important factors for adherence and are frequently the focus of health professionals.

But other issues relating to adherence abound. Health decision-making theories suggest that people must (a) be motivated to take their medications on a daily basis in order to be adherent to therapy, (b) be willing to prioritize treatment in order to persevere with a regimen over time, (c) have the confidence (both in oneself and in the health care system) to take medications, and (d) understand how treatment works, in order to contend with challenges such as side effects (Kalichman et al. 2001). Following medical advice properly is widely understood to be rational (Conrad 1985), and as described above, health behavior theories suppose that acts of ingesting medications are determined by values, beliefs (including information and education about illness and therapies), and expected outcomes (Munro et al. 2007). Hindrances to developing appropriate motivations, intentions, and beliefs are likewise conceptualized as being liable to increase nonadherence. This accounting of health behavior falls nicely in line with views of rational action: insofar as health decisions (such as pill-taking) are predicated on attitudes and beliefs of patients, the model presumes a fairly direct pathway from motivation to intention to action.

A substantial body of HIV adherence literature embraces the model described above, focusing on the mental functioning of patients (mood, substance use), patients' self-regarding attitudes (e.g., self-efficacy), and expectations about medications as factors relating to adherence. Medication beliefs (whether it does or doesn't work; what exactly it works for) and their importance to health are also associated with adherence to therapy in much of the HIV adherence literature (Boyle 2000; Horne et al. 2004). Pharmacists, practitioners, and the medical literature see patients as needing to be "coached" on expectations around medication issues so they understand the importance of taking medications in light of difficulties like pill size or side effects. The goal is to engender in patients a stable set of medically accurate beliefs about medications that are integrated in their larger value systems, enabling them to autonomously carry out the action of pill-taking.

This view of adherence is frequently repeated on hospital wards and in clinics, just as it is cited in antiretroviral adherence research. It is, for all intents and purposes, a "conventional" view of adherence, and it treats adherence as a form of rational action. Missy's health care providers were not convinced that she had the necessary combination of beliefs and desires concerning her medication regimen to be adherent, nor a willingness to change with respect to substance use (including a number of failed attempts at methadone maintenance therapy), which was also seen to interfere with developing adherence-related attitudes and behaviors. Her own request that staff administer her medications seemed to endorse a common assumption that drug users have particular difficulty with adherence to ART and that they require firm and structured interventions[1] in order to consistently take their pills. This story of adherence as rational action is a smooth story, whereby patient, medication, and health system factors are neatly delineated; adherence is a singular phenomenon. But as I discuss below, the seemingly clear-cut take on adherence to ART is in fact much messier in practice; the conventional, singular story requires the coordination of multiple enactments of adherence. Attending to the multiplicity reveals much about adherence and also about that which is implicated in the acts of coordinating.

ADHERENCE IN PRACTICE

Consider again the Nystatin that Missy hid in the garbage. As a prophylaxis medication to manage a potential opportunistic infection related to HIV/AIDS, it isn't, strictly speaking, part of the combination of medicines that make up highly active ART. Why should refusing that medication be framed as a form of nonadherence relevant to an antiretroviral regimen? It likely wouldn't have been captured by surveillance data on antiretroviral adherence had this been part of a clinical research study—here, prescription refill of *strictly* antiretroviral medicines would be considered. But by virtue of the medication being connected to her overall health in the context of HIV infection, as well as it being part of a recommended set of actions given to her by a physician, the disposal of this medication is taken, both by Missy and her health care providers, to be indicative of an overall picture of Missy's ability to adhere to her antiretroviral regimen.[2] Hence, she hid the unused vials.

The Nystatin story points to the ways in which adherence is understood differently in the different physical and rhetorical spaces where knowledge about it is generated. Within research practices, a relatively narrow set of parameters determines whether someone is part of the "nonadherent" group. Adherence measures rely on calculations based on prescription refill rates, direct data from electronic monitoring of pill vials, and patient self-reports. We can compare this to how the concept of adherence is mobilized in community clinic settings. For those who participate in directly observed therapy (DOT) programs, a medication administration record (MAR) marks a witnessed ingestion with a tick, a red "O" if the patient had to have their medications delivered by outreach services or an "x" for a missed dose. Those filling in the charts are fallible, and ticks may be missed or added in erroneously, resulting in idiosyncratic record keeping that is not accounted for in traditional health research methods (Elliott 2007, 24). Moreover, the individuals doing the ticking are rarely acknowledged as part of the action, when in fact the interactions between clinic staff and patients very much shapes adherence in the community. Within a clinical environment, the MAR may matter to the nurses and coordinators directly involved in DOT programs (e.g., in determining who needs an outreach visit, who might require further supports, who might not be a suitable candidate for the DOT program because of too many missed visits), but the MAR sheet is not necessarily the primary concern of physicians following individuals with HIV/AIDS, who mobilize a different set of beliefs and practices around adherence.

In contrast, encounters between physicians and patients predominantly measure adherence by way of viral load—if viral load is undetectable over a series of blood draws and it remains that way, a patient is believed to be adherent. Drug resistance testing provides another seemingly objective measure that is taken to pertain to adherence patterns over time. Testing often occurs when viral load is suboptimal, and a report codes various antiretrovirals in red, yellow, or green boxes based on viral genotype. As with a traffic light, green means "go": the medications listed in the green box are safe to use, as resistance mutations have not yet rendered the medication ineffective, those coded yellow have some resistance building, but may still be used, and those in the red box will no longer be effective. Nonadherence and the associated build-up of drug resistance mutations are presumed as medication options move from green to yellow to red boxes. Patient self-reports of adherence have relatively little utility, with the exception of reporting side effects from medications. That said, other pieces of information are relevant, offering a proxy measure of adherence for physicians within the clinic setting. These are generally much wider than proxy measures within research practices and may be relayed to medical staff by other clinic personnel, or witnessed by physicians directly: in the clinic, vaccinations and blood work are important, as is whether someone is coming in at the right time for their methadone prescription and seeing the same health care provider; if they are keeping weight on or not; if they are quietly sitting in the waiting area or causing disruptions with the administrative staff at the front desk. Following through with other forms of medical advice as well as general "good" behavior in clinical settings is deemed to have relevance for adherence to ART.

Field Notes, January 14, 2007

We're back in the hospital room after a long morning at the oncology clinic. One of Missy's nurses brings in her morning medications. She asks the nurse if these are her pills. "Horse pills, that's what I call them." Missy looks at the bunch of them in the disposable paper cup, but doesn't make a move to take them. She isn't sure what they're for. She opens the drawer to her night table and a single pill rolls silently around in the space. It says APO CAL on it. I see the pill and Missy says to me, "this is the first time I've been bad. I think it's my calcium pill, I didn't take it last night." I shrug my shoulders. She takes all of the morning meds. I give her the lone calcium pill saying, "here, might as well knock it back with the others."

The hospital has yet another set of concepts around adherence that are mobilized, some of which overlap with community clinics or research measures and some of which do not. On one level, adherence is taken for granted if the medication cup goes into the room full and comes out empty; paradoxically, it is taken for granted that the patient is nonadherent *overall*, if substance use is thought to be an ongoing issue while a patient is in hospital, even if they are taking their medications. When the women taking part in the study found themselves on the hospital ward for extended stays, they often felt as though they were being scrutinized by hospital staff and were considered nonadherent to care during the times they left the ward to go to the smoking balcony. Speaking to me about each woman's ability to remain hospitalized and in adherence to medical advice, I would be told by staff, "you know, she's been off the ward a lot," as a tacit indicator of the health care providers' lack of confidence in their ability to remain engaged in HIV/AIDS care. Whether or not they were taking all of their medications, or consuming their ARVs, yet tossing the Nystatin, the participants felt they were treated as though they were nonadherent to their HIV care and going to "use" each and every time they were out of eyesight.

Insofar as it is taken to be a reflection of nonengagement in the health care system, a reputation for leaving against medical advice ("AMA") offers further reflection on the ability to adhere to an ART regimen in the hospital setting. As leaving AMA is connected in the minds of health care professionals to substance abuse, there is a slippage between presumptions of drug use and presumptions of nonadherence, whether or not that is the case at any given time. Given that drug use is associated with nonadherence to ART in the medical literature, it is easy to understand how drug use becomes a *stand-in* for nonadherence. Being out on the hospital's outdoor balcony or having a past history of self-discharge on welfare day, therefore, serves as hospital-based indicators of nonadherence to ART.[3]

In many instances, health care providers working in outreach and home care settings offer yet another view on adherence. Actively using substances need not indicate nonadherence in these spaces, as providers were frequently confronted with drug paraphernalia and evidence of substance use, given that they were entering individuals' private domains to provide care. Instead, being able to have a conversation with somebody who was seen as hard to reach, having someone open their door to the worker and accept a cigarette or be taken to an appointment, were all indicative of "engaged" behaviors that were considered to bode well for adherence, even if substance use was ongoing. Similarly, one of the HIV home care patients I observed would sometimes skip certain medications during a particularly difficult week, but she would still understand herself as nonetheless adherent to medical care, insofar as she continued to welcome the nurses into her home, extend the usual courtesies, accept her diet supplementation, and possibly take certain other pills from the pillbox that week.[4] Her nurses would attempt to encourage her that if she was going to "cherry pick," she ought to take the antiretrovirals and leave behind the sleeping pills or psychiatric medications.

While the nurses still understood adherence to be about getting ART into the patient's body each and every day, their approach didn't take adherence to be about a narrow set of measures; it included an understanding of this particular patient's need to negotiate medication ingestion. In contrast to the antagonistic dynamics that frequently developed around presumed drug use in hospital settings, this approach to thinking about adherence was more likely to ensure that CD4 counts remain high and that other forms of pathology were promptly recognized. It also mapped more closely onto the views of adherence that patients themselves had.

MANAGING MULTIPLICITY

Despite the multiplicity within the spaces where adherence to ART is enacted, a singular, "commonsense" story concerning *what* constitutes adherence and *who* is adherent (or not) to HIV medications ends up being told. The story is this:

1. Adherence is about getting all of the combination ART into someone's body, greater than 90% of the time, in order to achieve viral suppression[5]
2. Health decision-making, including ingesting medications, is a form of rational action
3. Nonadherence to ART is a failure of rational agency; drug users who are consistently nonadherent on any number of fronts are nonagentic, due to certain types of illness (e.g., neurological sequelae of HIV/AIDS), mental capacity, and externalities such as addiction

In contrast to this tightly scripted view, the data I've discussed offers a decidedly more diverse day-to-day mobilization of adherence by various health care practitioners in the field and by patients themselves. This includes contradictions and uncertainties that are often left unexplored by those working in clinical settings as well as by much of the quantitative social science research concerning adherence to ART: that drug users could be adherent in *some* spaces, but drug use was seen as antithetical to adherence in other social locations; that various kinds of proxy measures were taken to reflect nonadherence, rightly or wrongly; that the very factors that made someone *want* to adhere to medication might also make it difficult to do so. How is it that the different meanings of "adherence" in various locations are amalgamated? How does this multiplicity become singular?

In order to achieve a relatively unified view of adherence and of drug users' ability to adhere to HIV medications within research settings, certain practices and inconsistencies need to be removed from the data sets, mashed into the right form, or ignored altogether; certain voices that speak to issues surrounding adherence to ART must be muffled, while others must ring loud and clear; some dimensions of adherence must be visible and some must attract decidedly less attention. Within quantitative studies, data are "cleaned"[6] in order to produce a consistent object of knowledge that can be studied and subjected to policy intervention within a biomedical reality. In a process of rationalization—whereby data is translated into a format appropriate for statistical analysis—the responses of a given study participant are divorced from one another, enabling the erasure of the kinds of contradictions that exist in everyday experience.

In the clinical setting, dynamic tensions around medications may be erased as part of the construction of a singular story about adherence. In my field site, more complex understandings of the challenges with medication were simplified by the presumption that medication use was rational, drug use was irrational, and any difficulties that drug users might have taking medications were likewise related to irrational behavior. Take Missy, for example, who once said to me: "I think about [HIV] all the time. I look at that hospital bed [nodding toward the bed in her room] and seriously, it's what reminds me, every day. These days, I'm not feeling so bad though. I'm taking the meds, that's helping."

Missy's life circumstances made it such that her private spaces are entwined with health services, such that illness, even when she's feeling a bit better, is ever present. Her nod toward the hospital bed was a gesture toward an underlying tension: her medications help with the symptoms of HIV infection itself, allowing her to feel better than she had previously, but they (along with the hospital bed) are also reminders of illness. Both she and the other participants would frequently discuss how reminders of illness could make it *more* difficult to stay engage with health services, while *at the same time* being a kind of motivation to receive care. Health behavior models premised on views of rational action would suggest that her knowledge of her illness and her understanding of how the medications are helping are what lead to adherence. But this singular story of adherence fails to address how her knowledge of her illness was *simultaneously* challenging for maintaining medication use, by virtue of the fact that HIV/AIDS treatment was also a painful reminder of perceived failings. When she struggled to adhere to treatment, her actions were chalked up to ongoing substance use issues. Instead of attending to the complexity surrounding meanings of HIV/AIDS care, Missy's nonadherence was seen as ongoing irrational behavior. Across a variety of locations, the problems generated by a continued sense of surveillance, lack of privacy, and the feeling of illness representing personal failure often translated into the sorts of actions that would be tagged as "nonadherent." These were rarely accounted for; substance use (real or presumed) was enough to explain away the complexities.

Norms of rationality are seamlessly interwoven into the practices that produce clinical experience and biomedical research; as both a heuristic device and a metaphor that shapes clinical interactions, models of health behavior informed by ideals of rational action enable the singular story to be told. This same rational actor framework provides the impetus for interventions to address nonadherence to ART, created by those deemed to have epistemic authority on the issue of HIV/AIDS care—medical practitioners and biomedical researchers. This, in turn, shapes subsequent interactions in the health care system that generate further adherence-related data. Consider again the way in which departures from care against medical advice were taken as indicative of nonadherence to ART in the hospital setting. As I have discussed elsewhere, this was often a self-fulfilling prophecy: presumptions of drug use and "bad" behavior on the part of patients tended, more likely than not, to escalate tensions between staff and patients, sometimes leading to the very circumstances (an angry departure against medical advice) that the staff were concerned about. This meant worsening health outcomes, missed methadone doses, and unfilled ARV prescriptions—forms of nonadherence conventionally measured through surveillance data (Berkhout 2014). Just as ongoing drug use is seen as contrary to rationally motivated behavior, models of adherence presume that following through with medical advice is rational. Populist understandings of drug use as irrational and irresponsible smooth the way to comprehending how drug use functions as a cause, correlate, and proxy measure of nonadherence, which in turn impacts interactions in practice, including the departures against medical advice that *do* ultimately impact adherence.

MULTIPLICITY AND EPISTEMOLOGY OF IGNORANCE

To this point, I have argued that in asking what adherence to ART *is* in HIV/AIDS care, a story of multiplicity emerges when we attend to actual practices in the field. This way of thinking about medical entities, as versions of an object that come into being through practices, is indebted to the philosophical and ethnographic work of Annemarie Mol. Foregrounding how objects are handled in practice, Mol argues that different enactments of a disease entail different ontologies and that a disease is made to cohere through various tactics (Mol 2002). I have taken up this line of thinking as it might apply to adherence in HIV/AIDS care. In some clinical and research spaces the meanings and practices of adherence neatly converge, and as an object of study, it looks like a singular phenomenon. However, when we attend to the day-to-day mobilization of adherence-related practices, the variations and differences become more apparent.

But having a singular story requires that some divergences and contradictions that exist in practice *not* be represented, or at least not be taken into the final accounting of what adherence is, who adheres to ART, and why or why not. As I have tried to demonstrate, understanding adherence as a singular entity requires an erasure of certain differences that exist in practice. In further exploration of this aspect of *how* singularity comes to be, and pushing at some of the boundaries of Mol's own work on multiplicity in medicine, I turn now to the epistemology of ignorance literature in order to explore the erasures that take us away from multiplicity within adherence. It is my contention that we can better understand how adherence to ART becomes a singular story by attending to the forces that produce ignorance concerning certain aspects of adherence.

In "Race and Epistemologies of Ignorance," Shannon Sullivan and Nancy Tuana describe ignorance as a substantive part of knowledge practices; in so doing, they draw particular attention to the ways in which ignorance about an object of knowledge is produced and sustained (Sullivan and Tuana 2007, 1–9). Ignorance requires effort to achieve, manage, and preserve; not merely reflective of *gaps* in knowledge, it is an active social production, even as some aspects of ignorance may not be consciously produced (Spelman 2007, 120; Bailey 2007, 77). This way of thinking moves beyond the enlightenment myth of ignorance as a kind of hollow space or dark cave into which the light of knowledge is pulled (Proctor 2008, 5). With these background assumptions in place, Linda Martín offers a broad typography of epistemologies of ignorance, giving an analysis of three different (yet interrelated) forms that ignorance might take. First, at the level of the individual, our general situatedness as knowers means that ignorance is a given for any epistemic situation. One may be unknowing with respect to certain inquiries, but not others, as epistemic advantages and disadvantages accrue in relation to specific kinds of inquiry (Martín 2008, 42–43). Second, epistemic ignorance can also follow from one's social location or group identity, which shapes (among other features) belief sets, practices of justification, as well as norms of coherence and plausibility. Group identities confer epistemic advantages or disadvantages *per se*, not only in relation to a given context of inquiry (ibid., 44–47). Finally, Martín considers what she refers to as an explicitly structural account of ignorance, in which cognitive dysfunctions are *more* than simply difference in group experience or expertise—they are part of the social fabric—and are a substantive practice that differentiates a dominant social group and distorts reality (ibid., 47–49).

For my purposes, I want to focus on the second and third forms that epistemic ignorance can take.

In my field site, the voices of individuals living with HIV/AIDS were frequently marginalized on the subject of adherence, particularly in comparison to HIV "experts" and clinicians, as well as in comparison to so-called objective technologies such as the MAR sheets, hospital charting, and blood work results (whether or not lab printouts actually included viral load). Women drug users were not taken to be paradigmatic of "good" epistemic agents. They were predominantly seen as irrational or untrustworthy. Their social identities marginalized them as knowers on a number of axes; they lacked standing to make claims based on their identity as women and as drug users. Widely shared (imaginative) conceptions posit drug users as unreliable, deceptive, and "chaotic," and these same descriptors frequently functioned in clinical settings. Drug users in the community were also depicted by health care workers as "childish," a label more profoundly applied to women. The infantilizing language and imagery commonly applied to the category serves to diminish credibility and rational authority, given that children are not thought to be able to achieve the standards of autonomous rational agency. As feminists have persuasively argued, historical representations of femininity have been linked to child-like behavior—frivolous, silly, head-in-clouds—so as to portray women as oppositional to the "man of reason" (Code 1991). For women in the study, these overlapping images reinforced their lack of epistemic standing. Moreover, most participants identified as Indigenous. The stereotypes that portrayed the participants as unreliable, manipulative, and childish are, I would argue, more profound for Indigenous women, who also face a colonial imaginary that depict them as being in need of a pastoral (i.e., "civilizing") power. As the women in the study navigated a hierarchical, predominantly Caucasian health system, epistemic authority was unevenly distributed on the basis of these intersecting social identities. What knowledge women living with HIV/AIDS had to offer about adherence to ART was only taken up insofar as it could be seamlessly integrated into the conventional story framed by norms of rational action.

Contradictory views of adherence to ART were taken as evidence that an individual was "difficult" or trying to manipulate the system, or it was simply brushed aside. Like Missy, Dee described her relationship with ART in contradictory articulations: that it was hard to remember to take pills when she didn't feel sick *and* that it was hard to follow through with care when she was *so* sick. With respect to the latter view, she described this as "digging myself a hole." This hole, importantly, was one that she was already in, because of the fact of her illness, which she understood to be part of the longer, challenging trajectory of her life—adoption by a White family, a difficult reunion with her birth family, and having to give up her own son for adoption. Not seeing a way out *because* of the illness also made it difficult to adhere to treatment. Institutional documents, clinical teaching, and research surrounding adherence legitimize only one aspect of her experience: it makes sense of the former statement that *not* feeling sick can mean a lack of motivation to take medications. The latter experience was sidelined when Dee interfaced with the health care system. In clinical settings the relationship between being very sick and nonadherent was seen as intricately tied to drug use, rather than to difficulties coping with serious illness.

The participants' knowledge concerning adherence to ART was marginalized on the basis of their social location as would-be knowers. As Tuana writes, ignorance is "linked to issues of cognitive authority, doubt, trust, silencing, and uncertainty" (Tuana 2008, 109). Moreover, presumptions of credibility, competence, and trustworthiness—presumptions underwriting epistemic authority—are distributed differentially according to the ascribed identities of speakers and interpreters (Code 1995). As knowledge concerning adherence to ART is smoothed into a singular story premised on rational action, the social identities of would-be knowers and a wider social imaginary surrounding instrumental rationality and rational agency help to determine those aspects of adherence that are seen and heard, and those that remain silenced. The complexities surrounding *who* is adherent versus nonadherent to ART and *why* are lost as knowledge. What results is the preservation of the conventional story (adherence to ART is a form rational action; nonadherence caused by failures of rationality), despite evidence to the contrary. The erasure of experiential knowledge due to a lack of epistemic standing is therefore implicated in the production of ignorance owing to social identity.

In a related vein, some outreach nurses providing home care to individuals living with HIV had experiences that demonstrated that nonadherence wasn't *strictly* about ongoing drug use, but these experiences were also treated as peripheral. Despite their experience that individuals actively using substances could be adherent with the right (albeit intensive) support, this belief was not shared among colleagues and higher ups. One outreach nurse described it as such:

> *A lot of the other nurses [who] see our patients, deep down, [they] feel that they should not be seeing them. So [the nurses] start off with an attitude that deep down, they're not happy about having to go [to see the patient] every day, and that colours how you react to people.*

Tropes attached to the group identity of patients/clients—of drug users as manipulative, childish, and difficult—functioned within the work site and were used by clinic managers and some coworkers to say that active substance use was evidence of being unmotivated to take medications and that limited health care resources would be better used elsewhere. Despite having some level of cognitive authority based on their status as professionals, the nurses who advocated for daily HIV care for substance users in the community faced serious pushback, to the point that they were at times treated like "outsiders" to the system. As outsiders, they were epistemically disadvantaged compared to their colleagues and managers.

We can understand this on the basis of the wider social imaginary that the nurses' knowledge claims came up against. Their knowledge of adherence pushed against the system of images, metaphors, and tacit assumptions surrounding rationality. That certain individuals who were using drugs *could* adhere when care was delivered in a respectful way was swept aside because it did not cohere with the "common" knowledge around substance use and adherence to ART, which I've argued is premised on a model of health behavior, itself informed by views of rational action. This is an instance of the third form of ignorance outlined by Martín: social scripts, cultural myths, and dominant views supply belief-inducing premises that are cast as unchallengeable and that persist even in light of counterevidence; they have an organizing role in the production of knowledge on a given issue (Martín 2007, 48–49). Norms of rationality and tropes of drug users are part of an overarching framework—an entrenched social imaginary—that sustains ignorance concerning adherence to ART and distorts reality such that counterevidence cannot be taken as such.

CONCLUDING THOUGHTS

Understanding adherence to ART means looking at more than why someone does or does not ingest a particular combination of pills. Adherence must be understood in light of the purposes and functions of those who mobilize the concept in specific locations. It must also be analyzed with regard to the function that the rhetoric of drug use, transgression, and deviance plays in the construction of knowledge concerning what it means to be nonadherent and concerning those who are thought to be predictably so. Rather than dichotomize adherence to ART as either an active engagement in services that are appreciated and seen as essential to survival or being not sufficiently motivated (due to substance use), I have articulated the dynamic tensions that play out in the acceptance of HIV care. On one level, this tells us more about what adherence *is*—or rather, how adherence is enacted in different locations, by different actors. Understanding the terrain of adherence in this more nuanced fashion is part of developing ways of engaging in health care practices that make room for complexity, as some of the outreach nurses were able to do. For HIV/AIDS programming, it means that large, scaled-up, "one-size-fits-all" interventions are unlikely to succeed if they cannot be adapted for a given place and time. Importantly, understanding multiplicity in adherence also asks us to measure outcomes in a manner that does not "clean" away data that can be hard to reconcile or contend with. There is also a larger point to be made here: medicine, write Mol and Berg, is a "heterogeneous coalition of ways of handling bodies, studying pictures, making numbers, conducting conversations" (Mol and Berg 1998, 3). When we attend to the practices and performances of medicine, we can see this multiplicity.

NOTES

1. In Vancouver's downtown eastside, this includes witnessed medication delivery programs such as directly observed therapy (DOT), a medication program modeled after structured tuberculosis interventions.

2. It is often the case that when patients who are seen as likely to have difficulty with nonadherence are going to begin ART, they are started with a two-week "trial" of vitamins, to see how they are going to do, whether they are able to take the vitamin each day, either at the pharmacy, at a directly observed therapy (DOT) program site, or via pharmacy bubble packaging, which is then examined by a nurse or physician to determine the level of adherence. "Engagement" regarding *other* forms of medical treatments is taken to have implications for antiretroviral therapy, which is why my analysis includes an interpretation of the conditions under which these diverse actions occur.

3. There is overlap between the hospital and the community clinic, inasmuch as discharge summaries and case conferences about behavior on the ward do sometimes make their way to clinical staff in the community. But for the most part, leaving AMA or being on the balcony are not indicators that are heavily relied upon within the clinic. Different proxy measures, also often related to assessment of frequency of drug use, are employed instead; these do not necessarily overlap with the proxies employed within adherence research.

4. Peter Conrad similarly describes this phenomenon, with respect to patients following treatment programs for epilepsy. Conrad writes, "Although many people failed to conform to their prescribed medication regimen, they did not define this conduct primarily as noncompliance with doctors' orders." Conrad conceptualizes this as indicative of "medication practices," in which the prescribed orders are just *one* part of a medication practice—the role (or lack of role) a medication plays in allowing one to live one's life "normally" involves many other medication and health-related considerations that individuals also conceptualized as relevant to their actions around medications, leading Conrad to conceptualize this as "self-regulation" of medications (as opposed to noncompliance with orders). See Conrad, 1985.

5. See, for example, The BC Centre for Excellence in HIV/AIDS (BC CfE) "Antiretroviral Update," which includes BC guidelines for treatment as well as a comparison between international guidelines and those suggested by the BC CfE.

6. "Cleaning" data is a term used by some researchers to depict the process of preparing data prior to analysis. It entails double-checking the coding system and resolving inconsistencies and discrepancies. It also includes ensuring that larger amounts of text are broken down into smaller categories (preferably binary numeric categories) that are meaningful for statistical analysis. The interesting thing, from my research, has been to see how inconsistencies in data are resolved. If a research participant responds to a question by saying "a few" (e.g., how many times a day do you inject heroin—"a few"), different individuals carrying out data entry might have a different understanding of how to represent this numerically. Likewise, there are inconsistencies in the interpretation of what someone might have meant when referring to entities not familiar to researchers entering data, if the individual who carried out the survey questionnaire is not available to respond, or does not remember that particular survey. It may also involve removing variables that are not thought to relate to analysis.

REFERENCES

Baily, Alison. 2007. "Strategic Ignorance." In *Race and Epistemologies of Ignorance.* Eds. Shannon Sullivan and Nancy Tuana, 77–94. Albany: State University of New York Press.

Bangsberg, David R. et al. 2003. "High Levels of Adherence Do Not Prevent Accumulation of HIV Drug Resistance Mutations." *AIDS* 17(13): 1925–1932.

Bartlett, John A., Ralph DeMasi, Joseph Quinn, Cory Cary Moxham, and Franck Rousseau. 2001. "Overview of the Effectiveness of Triple Combination Therapy in Antiretroviral-naive HIV-1 Infected Adults." *AIDS* 15(11): 1369.

BC Centre for Excellence in HIV/AIDS. 2008. *Antiretroviral Update.* Vancouver.

Berkhout, Suze G. 2014. "Bad Reputations: Memory, Corporeality, and the Limits of Hacking's Looping Effects." *PhaenEx* 9(2): 43–63.

Boyle, B. A. 2000. "HAART and Adherence." *AIDS Reader* 10: 392–396.

Code, Lorraine. 1995. *Rhetorical Spaces: Essays on Gendered Locations.* New York: Routledge.

Code, Lorraine. 1991. *What can She Know? Feminist Theory and the Construction of Knoweldge.* Ithaca, NY: Cornell University Press.

Conrad, Peter. 1985. "The Meaning of Medications: Another Look at Compliance." *Social Science and Medicine* 20(1): 29–37.

Elliott, Denielle. 2007. *Pharmaceutical Surveillance, Medical Research, and Biovalue among the Urban Poor.* Simon Fraser University.

Fricker, Miranda, 2007. *Epistemic Injustice.* Oxford; New York: Oxford University Press.

Horne, Robert, Deanna Buick, Martin Fisher, Heather Leake, Vanessa Cooper, and John Weinman. 2004. "Doubts about Necessity and Concerns about Adverse Effects: Identifying the Types of Beliefs that are Associated with Non-adherence to HAART." *International Journal of STD AIDS* 15(1): 38–44.

Kalichman, Seth et al. 2001. "HIV treatment Adherence in Women Living with HIV/ AIDS: Research Based on the Information-Motivation-Behavioural Skills Model of Health Behaviour." *Journal of the Association of Nurses in AIDS Care* 12(4): 58–67.

Law, John. 2004. *After Method.* 1st ed. New York: Routledge.

Martín, Linda. 2007. "Epistemologies of Ignorance: Three Types." In *Race and Epistemologies of Ignorance*, edited by. Shannon Sullivan and Nancy Tuana, 39–57. Albany: State University of New York Press.

Mol, Annemarie. 2000. "Pathology and the Clinic: An Ethnographic Presentation of Two Atheroscleroses." In *New Medical Technologies: Intersections of Inquiry*, edited by Margaret Lock, Allan Young, and Alberto Cambrosio, 82–102. New York, NY: Cambridge University Press.

Mol, Annemarie. 2002. *The Body Multiple: Ontology in Medical Practice.* Durham: Duke University Press.

Marc, Berg, and Annemarie Mol. "Difference in Medicine: An Introduction." In *Differences in Medicine*, Marc Berg, Annemarie Mol, Eds., p. 3. Copyright, 1998, Duke University Press. All rights reserved. Republished by permission of the copyright holder. www.dukeupress.edu

Munro, Salla, Simon Lewin, Tanya Swart, and Jimmy Volmink. 2007. "A Review of Health Behaviour Theories: How Useful are These for Developing Interventions to Promote Long-term Medication Adherence for TB and HIV/AIDS?" *BMC Public Health* 7(01): 104.

Osterberg, Lars, and Terrence Blaschke. 2005. "Adherence to Medication." *NEJM* 353: 487–97.

Proctor, Robert N. 2008. "Agnotology: A Missing Term to Describe the Cultural Production of Ignorance (and Its Study)." In *Agnotology: The Making and Unmaking of Ignorance*, edited by Robert N. Proctor and Londa Schiebinger, 1–33. Stanford: Stanford University Press.

Spelman, Elizabeth V. 2007. "Managing Ignorance." In *Race and Epistemologies of Ignorance*, edited by Shannon Sullivan and Nancy Tuana, 119–131. Albany: State University of New York Press.

Stone, Valerie, Joseph Hogan, Paula Schuman, Anne Rompalo, Andresa Howard, Christina Korkontzelou, and Dawn Smith. 2001. "Antiretroviral Regimen Complexity, Self-reported Adherence, and HIV Patients' Understanding of their Regimens: Survey of Women in the HER study." *Journal of Acquired Immune Deficiency Syndromes* 28(2): 124–131.

Sullivan, Shannon, and Tuana, Nancy. 2007. "Introduction." In *Race and Epistemologies of Ignorance*, edited by Shannon Sullivan and Nancy Tuana, 1–10. Albany: State University of New York Press.

Tuana, Nancy. 2008. "Coming to Understand: Orgasm and the Epistemology of Ignorance." In *Agnotology: The Making and Unmaking of Ignorance* edited by Robert N. Proctor and Londa Schiebinger, 108–145. Stanford: Stanford University Press.

World Health Organization. 2003. *Adherence to Long Term Therapies: Evidence for Action.* Geneva.

Chapter 13

Ebola and the Rhetoric of Medicine

Supportive Care and Cure

James Krueger

The WHO estimates that over 11,000 people were killed in the recent Ebola virus outbreak in West Africa.[1] For at least a short time the scope of this epidemic, unprecedented for this disease, captured the world's attention. This, in turn, reignited debates about the role wealthy countries should play in supporting health systems around the globe (see, for example, Rid and Emanuel 2014). Less obviously, however, the descriptions of Ebola virus disease that were widely offered also raised important conceptual questions within medicine. Ebola is often described as an untreatable illness. At the same time, however, medical providers within the affected nations, and from many organizations around the world, made heroic efforts in response to the epidemic. These efforts were not just limited to quarantine and infection control. They included direct care for patients. If these professionals were not offering treatment, what was it that they sought so urgently, at such risk to themselves, to provide?

One simple answer, and one that appears in many discussions of Ebola, is that they offered supportive care. This raises a question, then, about the meaning and scope of the term "treatment" and its relationship to such care. To add to the complexity, others have suggested not that the condition is untreatable, but that the only treatments available are not of the right sort. They are, for example, only supportive treatments, or nonspecific treatments. These different descriptions, then, raise related questions about the meaning and scope of these terms.

My aim here is to try to clarify some aspects of these descriptions of Ebola. This, importantly, is not a task of purely conceptual interest. One of the challenges in responding to outbreaks of Ebola, or of the closely related Marburg virus, is the fear they inspire. That fear is no doubt spurred by the painful symptoms and high mortality caused by these diseases. It is also influenced by the descriptions that are offered concerning the available medical response and the manner in which that response is carried out in affected communities. Given that fear can drive the spread of disease in an outbreak, it is crucial to be able to provide clear and accurate information about a condition and the interventions, if any, available to combat it.

In what follows, I will first summarize features of the medical response to filovirus diseases (encompassing both Ebola and Marburg). In characterizing this response, I will focus on the account of medical care provided by Médecins Sans Frontièrs (MSF) in the *Filovirus Haemorrhagic Fever Guideline* (hereafter, *FHF Guideline*). This will give information about the care that is made available in the context of most real cases of disease outbreaks.[2] The *FHF Guideline* also suggests that no curative treatment is available for Ebola or Marburg, describing the interventions within its pages as supportive treatments only. This distinction, between the curative and the supportive, will then be the central focus of the remainder of this chapter.

In trying to understand how this distinction is drawn, I will first turn to consider two general approaches. The first focuses on the intention of caregivers in offering care. The second considers whether an intervention addresses the cause of a condition, or just its symptoms. This second account seeks to connect the idea of a curative intervention with that of a specific intervention, regarding both as focused on underlying causes. I will argue that each of these accounts encounters signification difficulties. I will then develop an account that is meant to avoid the problems of these two proposals and close by highlighting the practical importance of paying close attention to such seemingly merely terminological issues.

FILOVIRUS CARE

Organizations such as MSF have played an important role in responding to outbreaks of filovirus disease. It is not surprising, then, the MSF publishes a comprehensive guide covering basic information about these diseases as well as detailed instructions for organizing a response. In characterizing the disease itself, the *FHF Guideline* notes that the symptoms are initially general in nature and "are similar to common tropical diseases like malaria, shigellosis, or typhoid" (Sterk 2008, 15). These general symptoms include such things as tiredness and weakness, nausea, and muscle pain. Later, symptoms develop to often include diarrhea, vomiting, rash, confusion and irritability, multiple organ failure, shock, and death (Sterk 2008, 15). Bleeding is generally found in less than half of the cases, a fact that has led the WHO to no longer refer to Ebola as a hemorrhagic fever, now preferring the more general Ebola virus disease (Quammen 2014, 47).

The part of the *FHF Guideline* of most interest here is dedicated to the description of available medical care. That section is divided into seven sub-sections. It begins with a directive about the proper washing of hands to avoid spreading infection, a warning followed by general observations about Ebola. These observation include the assertion that "currently there is no curative treatment for Ebola and Marburg Haemorrhagic Fever. Only supportive treatment can be offered to the patients" (Sterk 2008, 40). It continues by noting, "Experience in former outbreaks shows that supportive treatment reduces the suffering of patients and aggressive invasive supportive treatment might maximize chances of survival" (ibid.).

The remaining six sections on patient care are given topical headings. First is "Invasive procedures" followed by "Hydration," "Symptomatic care," "Presumptive treatment," "Supplementation," and finally "Nutritional support." The section on invasive procedures opens and closes with both a warning about the dangers such procedures present for medical staff and a reminder that "intensive supportive treatment," including invasive procedures, may positively impact patient outcomes (Sterk 2008, 41). The remainder of the section is given over to a listing of safety procedures that should be followed with any invasive procedure in order to minimize risks to caregivers.

The first discussion of any particular intervention comes with the section on hydration. In this very brief section, both oral and IV hydration are mentioned. Given the suggestion that aggressive, invasive support is regarded as improving outcomes, it would seem that IV hydration is preferred, though the *FHF Guideline* does not state this explicitly. It is also worth noting in this context that hydration is not grouped under the general heading of "Symptomatic care." That later section covers everything from recommendations for medications to reduce fever, to treatments for pain, nausea, anxiety, agitation, and confusion (Sterk 2008, 42–43). The separation of hydration from symptomatic care reinforces the suggestion that provision of fluids is of central importance. It is not merely another form of symptomatic care.

This conclusion is further confirmed by reference to lessons learned from the MSF mission during a Marburg virus outbreak in Uige, Angola. It is the report on medical care from this outbreak that is the reference supporting the assertion of a possible positive effect on patient survival from aggressive supportive care. The key passage from the report suggests that "although no strong evidence is available on benefits of iv fluids for patients with [Marburg hemorrhagic fever], and although the risks to staff cannot be denied, iv fluids were considered to be very helpful. Not only did they appear to improve survival, but they also appeared to greatly improve the patients' and their families' perceptions of the Marburg ward" (Jeffs et al. 2007, S158). Two cases in particular are mentioned as evidence of the benefit of IV fluids. In the first, a woman "survived 5 days unconscious despite a very hot climate," while in the second "a girl 5 years of age recovered from severe shock; both had confirmed [Marburg hemorrhagic fever] and received iv rehydration" (ibid.). The remaining sections of the *FHF Guideline* discuss treatment for common, endemic conditions that may be either mistaken for Ebola or constitute comorbidities requiring treatment ("Presumptive treatment"), recommended vitamin dosing ("Supplementation"), and procedures for feeding ("Nutritional support") (Sterk 2008, 43–44). This final section includes mention of the use of a nasogastric tube in some cases. Nasogastric feeding is also mentioned in the hospital report from Uige, primarily in the context of helping to convince local populations of the value of hospitalization. The report suggests that "proactive supportive treatment, including iv fluids and nasogastric feeding, may help persuade people to accept isolation, because patients often value the perceived effort" (Jeffs et al. 2007, S160).

In sum, the *FHF Guideline*, and related reports from the Uige Marburg outbreak, suggests that aggressive hydration in general, and likely IV hydration in particular, is of particular importance in the care of filovirus disease patients. The report from Uige singles out IV fluid as positively impacting impressions of patient survival. Circumstances may also warrant nasogastric feeding in specific cases, and a range of symptomatic and nutritional supports are indicated. At the same time, treatment for other common diseases (malaria, for example) is undertaken both because these other conditions can be mistaken for filovirus disease and because they may be examples of comorbidity. All of these interventions, in both the *FHF Guideline* and the report from Uige, are regarded to be supportive measures, not curative. Given that the *FHF Guidelines* clearly countenance support as a kind of treatment, those that refer to Ebola as untreatable are restricting the concept of treatment to what the *FHF* refers to as curative treatment. For the remainder of this chapter, I will treat supportive care and supportive treatment as equivalent, reflecting different ideas concerning the scope of the term "treatment" and not differences in the care itself. The question here, then, is what distinguishes supportive care from curative care.

INTENTION, SPECIFICITY, AND CURE

The phrase "supportive care" is not unfamiliar within medicine. Though most commonly occurring in oncology, it is not restricted to that context. It is often used interchangeably with other terms, such as "hospice care," "end of life care," or "palliative care," though there are recent efforts to try to disambiguate these concepts (see, for example, Cramp and Bennett 2013, 53–60). Those associations provide one starting place for thinking about the difference between curative interventions and supportive interventions.

The transition into a hospice program is usually marked by a shift in the aim of medical interventions. The measures that healthcare providers undertake are no longer aimed at curing a patient of his or her disease. Instead, the aim of any intervention is to maximize the available quality of life. Even among those seeking to clarify the nature of supportive care and distinguish it from hospice care, the emphasis on quality of life remains. Cramp and Bennett, for example, suggests that supportive care aims to "ensure the best possible quality of life" for patients throughout the course of their condition and that it should have "equal priority alongside diagnosis and treatment" (2013, 59). Hui et al. note that, based on their review of published definitions, "'supportive care,' 'palliative care,' and 'hospice care'" all "denote interprofessional care to optimize quality of life" (2013, 683). The main difference, on these accounts, between supportive care and hospice or palliative care is the association of the latter two with the end of life. Supportive care is thus taken to be broader in application, encompassing quality-of-life care at all stages of disease (see, for example, Hui 2014). This suggests that one possible difference between curative treatment and supportive treatment is the aim or intention of the care. If the intent in offering the care is to improve quality of life, the care is supportive. If it is to combat and/or remove disease, it is curative. We might, then, define terms in this way:

Curative$_1$ Intervention: Intervention primarily intended to restore a patient to health.
Supportive$_1$ Intervention: Intervention primarily intended to maximize available quality of life.

Importantly, these definitions leave open the possibility that supportive interventions can positively impact survival and recovery. What matters is the primary intent of the care, not its ultimate effects. Thus, while management of side effects from chemotherapy can be aimed at making a patient more comfortable, increased strength and comfort may also improve a patient's ability to withstand therapy. In a similar way it would be possible, in cases of Ebola, for rehydration therapy to positively impact outcomes, while still counting as supportive, so long as the intent of the care remains focused on comfort (quality of what life remains).

It can be difficult, however, to assess the intent of a particular intervention, especially when it is acknowledged that it might have further effects on overall outcomes. Remember that in the *FHF Guideline*, hydration is given its own section, separate from "Symptomatic care." Aggressive supportive care is particularly singled out for its potential impact on patient survival. Crucially, the *FHF Guideline* concludes its discussion of invasive procedures (including IV fluids) with the following, highlighted paragraph:

Each invasive procedure is a dangerous action for the person performing the procedure and his assistant. Therefore limit the invasive procedures to the absolutely necessary, but keep in mind that intensive supportive treatment may have a positive impact on the outcome. (Sterk 2008, 41)

Invasive procedures carry the real risk of caregivers contracting the disease. This warning, however, is balanced against the repeated emphasis in the *FHF Guideline* that such measures potentially decrease patient mortality. It is the chance to save lives that gives a reason why such procedures might be attempted. Given the risks, if it could be shown that IV fluids had no impact on patient survival, there would seem to be little justification for undertaking such care. Other measures, pain relief, oral rehydration, measures to control anxiety, and so on, would surely still be warranted because they do not pose the same level of risk. But, in undertaking invasive procedures, the aim seems clear. It is to maximize the chance of patient survival and recovery. It is an attempt to help restore the patient to health. Thus, the administration of IV fluids doesn't seem to primarily aim at maintaining quality of life.

If we abandon an account focused on intention, one alternative to consider would focus on differences in the nature of the treatments themselves. Returning to the *FHF Guideline*, for example, we can note an additional use of the term "supportive treatment"; it is, at least in part, an expression of the limitation of current treatment options. This suggests there is something about the treatments themselves that is different, not just the intent driving the care. Pointing out the lack of curative interventions is not a plea for caregivers to shift their thinking about the care that is available; it is an expression of the need for new treatment modalities. While sometimes supportive care might be distinguished by intention (as in differentiating radiation for palliation from radiation for cure), at other times the nature of the treatment itself might be such that it is supportive, regardless of the intentions of the caregiver.

What, then, might differentiate the interventions discussed in the *FHF Guideline* from curative interventions? What is missing in the case of the procedures described? Here, it is worth noting that in each case, including rehydration, the intervention fails to target the virus itself. None of the interventions can inhibit the ability of the virus to infect cells, to replicate, or to survive in the body. Thus, the central cause of the condition is not affected by those measures that can be taken. This, then, might provide a basis for the distinction. Curative treatments target the cause of a disease, while supportive measures do not. In this way, such an account would depend on at least two things. It is tied, first, to an understanding of the central causal feature (or features) of a condition and, second, to an understanding of the mode of action for an intervention. If an intervention can be shown to directly target the central cause of disease, then it is curative. If it does not, it is, at best, supportive.

One advantage of such an account is that it provides a way to unify some of the diverse ways that Ebola is discussed. In particular, it links the definition of a curative treatment with at least one possible understanding of "specific treatment." A treatment might be deemed specific precisely because it is targeted at the cause of a disease. It has specific application to this causal agent, in our example the Ebola virus or Marburg virus. Supportive care is thus not specific for the same reason it is not curative. It is not targeted in this way. It is important to note that this is not the only possible understanding of "specific treatment," but it is one well rooted in medical usage. For example, *Stedman's Medical Dictionary* (2006) defines "specific therapy" as "therapy aimed at the cause(s) of a disease process, as opposed to symptomatic therapy." I will consider an alternative account of specific therapy later.

This understanding of the difference between supportive and curative treatment parallels an account developed by Stegenga. He suggests that an effective medical treatment is "one that improves the health of patients by curing disease or at least treating the symptoms of disease" (Stegenga 2015, 34). He then links this understanding of effectiveness with an understanding of what constitutes pathology. In his view, disease involves "both a biological component and a normative evaluation of that biological component" (35). In other words, the difference between health and disease is marked by real biological differences, but not just any differences will do. Those differences must have some negative impact on the organism in question. This approach allows him to distinguish two possible targets for an intervention, what he terms "causal target of effectiveness" and "normative target of effectiveness." The former directly addresses the relevant causal biological features of the disease, while the latter mitigates the negative effects. Thus, as he points out, pain relievers generally address only the symptoms of a condition (satisfying normative target of effectiveness), while antibiotics aim to eliminate the cause of a disease (satisfying causal target of effectiveness) (40).

These two targets, then, give an underpinning for the difference between cure and care. He asserts,

> If a medical intervention satisfies *causal target of effectiveness* then the intervention can be used to *cure* (or at least mitigate the constitutive causal basis of) a disease. If a medical intervention satisfies *normative target of effectiveness* then the intervention can be used to *care* for a patient with disease. (40)

This approach nicely proposes a reason why a specific treatment, one that targets the causal target of effectiveness, is preferred to a supportive treatment. On this reading, supportive care "modulate[s] only the symptoms of a disease without modulating the constitutive causal basis of a disease" (40). As valuable as this is, it is not the most fundamental aim of medicine "because cure typically also offers care, but not vice versa" (40). Supportive care can be valuable and effective at alleviating the symptoms of disease, but it cannot affect a cure because it does not affect the fundamental cause of a condition.

This account also nicely underscores that it is not just the target that matters; it is important that the intervention is effective, that it actually does cure disease or alleviate symptoms. This means that just having some effect at some level is not sufficient. In making this case, Stegenga notes several examples, including drug therapies aimed at multiple sclerosis that affect certain "biomarkers" of the disease while having "little impact on phenomenological (patient-relevant) symptoms" (42).

With this, we might arrive at the following definitions:

Curative$_2$ Intervention: An intervention that affects the causal basis of a disease in such a way as to remove or mitigate that basis.

Supportive$_2$ Intervention: An intervention that alleviates the symptoms of a disease without affecting its causal basis.

At first glance, these definitions seem to fit well the example of hydration and Ebola. Hydration doesn't target the causal basis of disease, the action of the virus on the body. It does, however, aim to prevent one of the important negative effects that viral action has, dehydration brought about by diarrhea and vomiting. As valuable as this may be, the suggestion is, it will not cure the disease.

There are, however, cases that raise questions about this basic approach. Consider, for example, the case of cholera. Cholera is a particularly telling example in comparison to Ebola because the acknowledged standard of care in cholera cases is rehydration therapy alone. MSF also publishes guidelines for cholera response. The outlined program for case management asserts that "most patients can be treated using oral rehydration solution (ORS) alone" (Baurenfeind et al. 2004, 45). It goes on to give detailed instructions for monitoring and assessing rehydration for patients with varying degrees of dehydration. Long experience treating cholera patients has provided evidence to support specific recommendations that are clearly not available for Ebola or Marburg.[3]

Importantly, however, for the discussion here, the MSF *Cholera Guidelines* goes on to consider the necessity of antibiotic therapy. This discussion begins with the following remarks:

In severe cases, antibiotics can reduce the volume of diarrhea and carriage time of *Vibrio*, but they are known to induce a false sense of security, leading to underestimation of rehydration needs. *Most cholera patients are cured by rehydration and do not need antibiotics.* On the other hand, if not correctly rehydrated, patients will die even if antibiotics are given. (Baurenfeind et al. 2004, 51, emphasis added)

There is no hesitation in referring to rehydration therapy as curing cholera. While the WHO fact sheet on cholera does not use the language of cure, it does assert that "up to 80% of cases can be successfully treated with oral rehydration salts" and moreover, argues that "mass administration of anti-biotics is not recommended, as it has no effect on the spread of cholera and contributes to increasing antimicrobial resistance."[4]

If curative treatment is defined as an effective specific treatment, effective treatment targeted at the cause of a disease, then rehydration therapy is not specific for cholera, and thus, it is not a cure for cholera. It does not affect the bacterium that is the cause of the disease. Instead, it prevents one major effect, severe diarrhea, from causing patient death. And yet, as the *Cholera Guidelines* points out, administration of antibiotics alone will not cure patients in the absence of rehydration. Administration of rehydration salts will, in the vast majority of cases, result in a patient recovering from the disease without the use of any antibiotics. The supportive care, targeting only the negative effects of the underlying disease, is an effective way to restore health. It alone saves the lives of most patients suffering from cholera.[5]

One might object that the immune system plays the crucial, curative role in this example. The causal basis of the condition, the bacteria, is still directly addressed, just not by the medical intervention. What hydration does is allow the cure, the immune system, time to work. It is, then, perhaps still a support-ive therapy and not curative. This would mean that discussions of cholera, to be technically precise, should not suggest that rehydration cures the disease. The fact remains, however, that administration of rehydration salts will result in dramatic reductions in mortality from cholera. Both the WHO and MSF go so far as to *discourage* the use of the "curative" treatment, antibiotics, in favor of what would be classified as supportive measures alone. The account of cure, here, begins to sound strikingly counterintuitive. It does not matter, in the end, how an intervention brings about a dramatic improvement in sur-vival from an often-fatal disease (whether directly affecting the bacteria or allowing the immune system to do its work). What matters is that it does. If it does, it is curative.

However, even if we were to bite this bullet and accept the implication that rehydration should not be regarded as a cure for cholera, there is still a diffi-culty. It would now be true that, at least in some cases, supportive treatment is *preferable* to curative treatment. This possibility highlights a deeper difficulty for applying the account we are considering to descriptions of Ebola. The lack of a curative treatment is lamented in the case of Ebola. Since supportive treatments clearly are available, reports that the condition is "untreatable" have to be understood as asserting a lack of curative treatments. If, however, a curative treatment is defined in the way we are considering at the moment, there need not be a preference for a specific, curative treatment. As we have seen, a supportive treatment could be preferred in cases where it is effective at bringing about patient recovery. The preference for curative treatment, then, depends on a judgment about overall effectiveness, not on anything about the nature of curative or supportive (or specific) interventions. The curative/supportive distinction, on this reading, is beside the point and misleads more than it informs.[6] Thus, this understanding of the distinction between supportive and curative (specific) treatment struggles to make sense of the central reason these concepts are being introduced in discussions of Ebola. Without providing a conceptual basis for lamenting the lack of curative interventions, the account is not going to prove adequate for understanding the very reason the distinction is drawn.

Perhaps, however, there is a way to avoid this conclusion, and at the same time render consistent the description of rehydration as curative in the case of cholera and supportive in the case of Ebola. Consider two possibilities. First, perhaps we could regard hydration as affecting the causal basis of disease in the case of cholera. The striking effectiveness of the intervention is due to the ability of the intervention to allow the immune system to clear the bacterium. In this sense, it does target the causal basis of the disease. If we attempt to go this route, however, it becomes unclear why hydration therapy cannot have the same kind of effect in the case of Ebola. We don't have conclusive evidence that it does, but that fact doesn't warrant the claim that such an intervention is supportive and not curative. At best, it supports a weaker, epistemic claim that we do not know if rehydration is curative. I will return to this possibility in a moment.

A second approach would begin with the observation that defining curative treatment in terms of an intervention's effective targeting of the cause of a disease requires some understanding of the relevant causal relationships. The classification of a treatment is then sensitive to shifts in the relevant casual understanding. At the same time, those causal structures depend on what we understand to be the pathology in each case. For example, the WHO reports that "80% of persons infected with the *V. cholera* do not develop any symptoms."[7] Thus, one might argue that from the point of view of therapeutic medicine, diarrhea is truly central to the disease and perhaps event constitutes the clinical disease itself, not the infection.[8] While the infection is still a causal factor, of more clinical importance is managing the effects of the diarrhea. Diarrhea is the central cause of the clinically important pathology. As we have seen, the *Cholera Guidelines* emphasizes that "if not correctly rehydrated, patients will die even if antibiotics are given" (Barenfeind et al. 2014, 51). In this way, one might argue that treating the effects of diarrhea *is* specific, curative treatment in the relevant sense. It treats the direct cause of mortality for this disease. Mortality is the central symptom, and diarrhea the primary cause.

There are two problems with this response. First, it does seem important to know if the disease is cholera (and amenable to rehydration alone) or some other disease that might require a different response. In this sense, while diarrhea is clinically relevant, surely it matters that *Vibrio cholera* is the cause of that symptom. Second, returning to discussion of Ebola, we can note that even if such a move were available for cholera, it is hard to see why it would be ruled out for Ebola, given what we currently know about the disease. On this understanding, assertions that no curative treatments are available, or no specific treatments are available, would seem to imply specific claims about the causal structure of Ebola. They would have to depend on the claim that the dimensions of Ebola virus disease treated by aggressive IV fluids are not of central importance to the clinical presentation of that disease in the way that diarrhea is central to cholera. In other words, these claims are no longer just assertions about the kinds of treatments available; they are claims about the details of how the Ebola virus affects the body. More specifically they are claims about what aspects of its action are of particular relevance to clinical practice. Such a specific claim about how we should understand Ebola virus disease, however, seems far removed from what is intended by such assertions. It would be odd, to say the least, to read "Ebola is currently untreatable" or "only supportive care can be offered for Ebola," as suggesting that diarrhea and fluid loss are not central clinical features of the disease.

AN ALTERNATIVE ACCOUNT

Throughout the arguments above, one of the recurring concerns has been whether an account of supportive care can sustain the sense that such care is lacking something important, that it is somehow not adequate in itself. The observation that IV fluids might improve patient survival, despite being classified as supportive, makes this no straightforward task. As we have seen, attempts to focus on either the intention in offering care, or features of the target for such care, struggle to account for this feature of discussions of Ebola. Either the care doesn't well fit the understanding of supportive care, or it fails to identify any necessary reason to prefer a "curative" intervention.

This, however, seems a very counterintuitive result. Surely curative interventions are to be preferred to merely supportive. Given that IV fluids in particular have been the source of much of the difficulty, a different way forward would be to accept that this intervention might ultimately prove to be curative if it proves truly effective at lowering mortality.

At the same time, at least some have asserted that hydration can be specific care as well. Consider, for example, Lamontagne et al.'s assertion that "supportive care is specific care for [Ebola virus disease]—and in all likelihood reduces mortality" (2014, 1565). In saying this, they are directly challenging the assertion that there is no specific intervention available for Ebola, while leaving open questions about effectiveness. What, then, is specific care in their estimation? Their description of Ebola virus disease emphasizes that after first presenting with general signs such as "fever, asthenia, and body aches" within relatively short order "the predominant clinical syndrome is a severe gastrointestinal illness with vomiting and diarrhea" (ibid.). Given that we know how to respond to the threat posed by such a clinical syndrome, we have specific information about what kind of care to provide. As they point out, with conditions that can cause intensive fluid loss, "when patients can no longer drink, placement of an intravenous catheter and delivery of appropriate replacement solutions are required" (ibid., 1566). This isn't a claim about Ebola *per se*, but a statement about the normal, expected care in the face of a disease that is having certain clear effects. In other words, in ways analogous to cholera, the clinical syndrome that presents itself is clear and the medically appropriate responses are well known. However, they report having seen "many critically ill patients die without adequate intravenous fluid resuscitation" during their work responding to the recent Ebola outbreak in West Africa (ibid.).

This suggests a different account of specificity than the one considered earlier. Rather than being specific to a causative agent, the care is specific to the "syndrome," gastrointestinal illness with vomiting and diarrhea, that presents itself clinically. In this sense, it is similar to the attempt to revise that causal account of specificity considered above, though it isn't making any claims about relative place in the causal structure of the disease as we understand it. It is enough to know that the patient is presenting with symptoms that are likely to lead to extreme fluid loss and that they then require aggressive therapy in response. Insofar as this is a part of the clinical presentation of the disease, response to it is "specific" to the disease regardless of how "central" such symptoms are to the causal basis of the disease. It may still turn out to be the case that rehydration is not effective, or is very limited in its effectiveness, at improving patient outcomes. Other effects of the underlying cause may continue to bring about patient deaths even in the face of adequate rehydration. On this understanding of specific care, however, this would not undermine the specificity of the response to the clinical syndrome. The care is still specific, even if it is not curative.

What about the distinction that has been our central focus, that between curative treatment and supportive treatment? Here, Lamontagne et al. seem to accept that IV fluids are a form of supportive care. They assert that they believe "we can and must do better at providing supportive care" and specifically mention the placement of IVs as an illustration (ibid.). As we have seen, it is difficult to see how to distinguish supportive care from curative care if supportive care can, like IV fluids, potentially have a direct impact on patient survival. Instead of accepting that IV fluids *are* supportive care, we can instead suggest that they are, *at worst*, supportive care (but may be more). One thing the idea of syndrome-specific care highlights is that a disease can have multiple different effects. Care can be adequate and specific for one effect of the condition, but death can still result because of other impacts of the disease. What we know in the case of cholera, to return to that comparison, is that rehydration alone is highly effective. Moreover, we have good evidence concerning how effective it is (what percentage of cases respond positively to oral rehydration alone) and what specific levels of care to provide depending on the manifestation of the disease in a particular case. In this way, the *Cholera Guidelines* provides much more detail about dosing of rehydration solution than the *FHF Guideline*.[9]

Lacking clear and convincing evidence that rehydration is effective in the same way for Ebola virus disease, it would seem premature to assert that such therapy is curative. At the same time, the syndrome-specific nature of hydration assures us that it is important to care for this disease. If we don't treat the effects of dehydration, patients will die. As in the case of cholera, curative interventions directly impact patient mortality (at least in potentially lethal conditions). Supportive interventions, then, at best impact mortality when combined with a primary, curative intervention. They are not sufficient on their own. This suggests the following definitions:

Curative$_3$ Intervention: An intervention that, on its own, is able to reduce patient mortality.

Supportive$_3$ Intervention: An intervention that either alleviates the symptoms of a disease or improves patient mortality when applied in conjunction with a curative intervention.

These definitions are, clearly, limited in application. They are tied to cases of potentially fatal conditions and likely to nonchronic conditions as well. Further analysis would be needed to extend or revise the account to fit further cases. In the context of Ebola and Marburg, however, they allow us to avoid the problems considered above. What is lamented, on this account, is the lack of a clearly established curative intervention, not some different treatment modality. It may be that no curative treatment is currently available, if the anecdotal evidence in support of IV fluids in particular turn out not to hold up under further scrutiny. But it is possible that such therapy is curative. In this sense, the lament is best focused on the lack of evidence, not the nature of the care. At a minimum, the syndrome-specific nature of rehydration gives us good reason to believe it is supportive. That it will, again at a minimum, play a role in maximizing the chance of recovery if and when genuine curative treatments become available. We can say this because we know the risks of dehydration, and we know generally effective ways to respond. Again, what we do not yet know is if this case is similar to cholera (or the degree to which it is similar), where such care is indeed curative.

CONCLUSION: FEAR AND EBOLA CARE

One of the striking features of Ebola is the fear that it engenders. Media reports concerning Ebola, not surprisingly, emphasize the lack of available treatments for the disease. These reports have fueled a wide range of responses, including, for example, attempts to quarantine healthcare professionals returning to the United States after working in West Africa during the most recent outbreak. At the same time, in West Africa itself, fears of the disease, along with rumors of a conspiracy involving foreign healthcare workers, led to attacks on treatment facilities, something seen in earlier filovirus outbreaks as well.

Such attacks have forced organizations such as MSF to work hard to find effective ways to engage with local communities during an outbreak. For example, a second report from the 2005 Marburg epidemic in Uige, Angola, identifies lessons learned from efforts to work with the local population. The report explains:

> The initial message about [Marburg hemorrhagic fever] relayed to the community included the statement that "There is no cure for this disease." This was understood by community members to mean "Even if I accept hospitalization, death is certain." Acceptance of isolation, thus, had to rely on the entirely altruistic motive of protecting one's family and neighbors from infection. Not surprisingly, some indigenous community-based healers promised a cure and distracted the community from following the advice of the response team. Later, in an attempt to correct this "alarmist" message, the information, education, and communication (IEC) program ... emphasized that infected individuals should come to the hospital and receive treatment. However, the message was somewhat overcorrected and now made the optimistic claim that "patients will survive due to hospital treatment," as opposed to "patients have a better chance of surviving if treated at the hospital," with the risk of nurturing unrealistic hopes. This illustrates that disseminating accurate and realistic message sometimes means walking a fine line. (Roddy et al. 2007, S166)

The authors of this report concluded that "information, education, and communication" efforts "contributed to nonacceptance of the Marburg ward and security problems in the community" in large part by "reinforcing fear and despair" (ibid.). They recommend that in the future, response teams "should emphasize that treatment—albeit limited in effectiveness—and care from medical professionals is available at the [filovirus hemorrhagic fever] ward" (ibid.). This example highlights the practical importance of clear terminology and the complex relationship between ideas such as "treatment," "cure," and "care" both within local communities and among medical professionals. Claims about a lack of cure were replaced with positive assertions about treatment that needed to be tempered to avoid creating damagingly unrealistic expectations.

One aim, then, of the discussion here has been to try to arrive at some clarity concerning the state of our response to Ebola. There are practical considerations to be kept in mind as well. Any description is going to need to be clear and easily communicated to diverse populations. Here, the arguments offered reinforce some of the practical lessons drawn by the MSF team. While it is certainly true that we do not know how effective the interventions we have available are, and it may still turn out to be the case that they have limited impact on patient survival, there is no conceptual reason to regard them as necessarily ineffective or as essentially divorced from the possibility of cure. We also do not know how limited in effectiveness aggressive care is with currently available techniques. There are real dangers of spreading false hope, but also risks in a failure to acknowledge the possibility of hope at hand.

Finally, it is worth emphasizing that none of the arguments here should be construed as suggesting there is no need to seek new therapies. Any treatment that could be demonstrated to be effective at reducing mortality would be enthusiastically welcomed by all (as has the recent news of a promising vaccine). But it also means that one, perhaps overlooked, route to identifying a curative treatment is the search for better evidence in the case of care that is already available. There are things that we know about responding to clinical syndromes even in the absence of detailed biological knowledge. This does not guarantee an effective response, but it does give a basis for hope that we can positively impact patient outcomes. Or, to borrow the closing of Lamontagne et al.'s plea for improved clinical care to combat Ebola (quoting William Osler), one route forward can be found in "do[ing] today's work superbly well" (Lamontagne et al. 2014, 1566).

ACKNOWLEDGMENTS

Thanks are owed to Oliver Brown for assistance with research for this chapter, and to Jacob Stegenga, Robyn Bluhm, and Elizabeth Clark for helpful comments and suggestions.

NOTES

1. World Health Organization, *Ebola Virus Disease Outbreak*, accessed January 20, 2016, http://www.who.int/csr/disease/ebola/en/.

2. Focusing on the care available in these contexts also limits some of the interventions that will be considered. I will not discuss, for example, the use of interferon or transfusions from patients who have recovered from the disease because they are not widely available options in an outbreak. Importantly, as should be clear later, the inclusion of such interventions would only strengthen the arguments offered here. These interventions are both more directly aimed at restoring health and more directly targeted at the cause of disease.

3. This also hints at a second, possible meaning of "specific treatment" that I will consider below.

4. World Health Organization, *Cholera Fact Sheet*, accessed January 20, 2016, http://www.who.int/mediacentre/factsheets/fs107/en/.

5. The *Cholera Guidelines* estimates that, without proper case management, up to 50% mortality rates can occur, while with proper treatment that rate can be less than 2% (Baurenfeind et al. 2004, 14).

6. This is true even if cholera turns out to be a relatively unique example. The point here is that the use of the curative/supportive distinction in this way distracts attention from the real central question, that of effectiveness. It does not matter what the target of the intervention is (the basis of the distinction on the reading being considered at the moment). What really matters is how effective it is at increasing chances for recovery. The use of the term "curative" in the way under consideration here is misleading insofar as it suggests greater effectiveness than mere "support" when in fact it is possible for supportive measure to be more effective interventions.

7. World Health Organization, *Cholera Fact Sheet*, accessed January 20, 2016, http://www.who.int/mediacentre/factsheets/fs107/en/.

8. The same would not be true from a public health perspective, where preventing the spread of the bacterium is of fundamental concern.

9. This is, perhaps, a further sense of specificity that could be distinguished from that just considered. This lack of specificity in the treatment guidelines for Ebola, beyond the general recognition of rehydration as a response to presentation of severe diarrhea and vomiting, may account for some asserting that we lack specific treatment for Ebola. On this reading, what those claims mean is we lack specific treatment guidelines that have been shown to be effective if followed. These two accounts share a crucial connection, as we will see more clearly in a moment, to the need for demonstrated effectiveness.

REFERENCES

Baurenfeind, Ariane, Alice Croisier, Jean-Francois Fesselet, Michel van Herp, Elisabeth Le Saoût, Jean McCluskey, and Welmoet Tuynman. 2004. *Cholera Guidelines, Second Edition.* Médecins Sans Frontiéres.

Cramp, Fiona, and Michael Bennett. 2013. "Development of a Generic Working Definition of 'Supportive Care'." *BMJ Supportive and Palliative Care* 3: 53–60.

Hui, David. 2014. "Definition of Supportive Care: Does the Semantic Matter?" *Current Opinion in Oncology* 26: 372–9.

Hui, David, Maxine De La Cruz, Masanori Mori, Henrique A. Parsons, Jung Hye Kwon, Isabel Torres-Vigil, Sun Hyun Kim, Rony Dev, Ronald Hutchens, Christina Liem, Duck-Hee Kang, and Eduardo Bruera. 2013. "Concepts and Definitions for 'Supportive Care,' 'Best Supportive Care,' 'Palliative Care,' and 'Hospice Care' in Published Literature, Dictionaries and Textbooks." *Supportive Care in Cancer* 21659–85.

Jeffs, Benjamin, Paul Roddy, David Weatherill, Olimpia de la Rosa, Claire Dorion, Marta Iscia, Isabel Grovac, Pedro Pablo Palma, Luis Villa, Oscar Bernal, Josepha Rodriguez-Martinez, Barbara Barcelo, Diana Pou, and Mattias Borchert. 2007. "The Médecins Sans Frontièrs Intervention in the Marburg Hemorrhagic Fever Epidemic, Uige, Angola, 2005. I. Lessons Learned in the Hospital." *The Journal of Infectious Disease* 196: S1584–161.

Lamontagne, François, Christophe Clément, Thomas Fletcher, Shevin T. Jacob, William A. Fischer II, and Robert A. Fowler. 2014. "Doing Today's Work Superbly Well – Treating Ebola with Current Tools." *New England Journal of Medicine*, 371: 1565–6.

Quamman, David. 2014. *Ebola: The Natural and Human History of a Deadly Virus.* New York: W. W. Norton and Co.

Rid, Annette, and Ezekiel Emanuel. 2014. "Why Should High-Income Countries Help Combat Ebola?" *JAMA* 312: 1297–8.

Roddy, Paul, David Weatherill, Benjamin Jeffs, Zohra Abaakouk, Claire Dorion, Josefa Rordiguez-Martinez, Pedro Pablo Palma, Olimpia de la Rosa, Luis Villa, Isabel Grovas, and Matthias Borchert. 2007. "The Médecins Sans Frontièrs Intervention in the Marburg Hemorrhagic Fever Epidemic, Uige, Angola, 2005. II. Lessons Learned in the Community." *The Journal of Infectious Disease* 196: S162–7.

Stedman's Medical Dictionary. 2006. Philadelphia PA: Lippincott Williams & Wilkins.

Stegenga, Jacob. 2015. "Effectiveness of Medical Interventions." *Studies in History and Philosophy of Biological and Biomedical Sciences* 54: 34–44.

Sterk, Esther. 2008. *Filovirus Haemorrhagic Fever Guideline.* Médecins Sans Frontiéres.

Chapter 14

Action, Practice, and Reflection

Dewey's Pragmatist Philosophy and the Current Healthcare Simulation Movement

Joseph S. Goode

In 1999 the Institute of Medicine (IOM) released a report entitled *To Err Is Human*, which outlined the then current understanding of error genesis in modern healthcare practices as well as error prevalence and incidence. The IOM estimated that 100,000 fatalities per year could be directly related to medical errors. As it turns out, this number was likely a significant underestimation, but the impact of the report was profound. Patient safety moved front and center in both the professional and popular literature, and efforts such as the 100,000 Lives Campaign were initiated. One other consequence of the IOM report was an exponential growth in the use of simulation in healthcare education. Drawing from the model in commercial and military aviation, healthcare simulation (HCS) was initially a small niche in the specialty area of anesthesia. The IOM specifically cited simulation as a means to reduce error in healthcare processes, and subsequently, interest in the use of this educational modality accelerated. A marker of this increased interest and usage is the volume of publication in the area. Issenberg and McGaghie (2005) identified 670 articles related to medical simulation published between 1969 and 2003. In 2011, Cook et al. identified a pool of 10,903 articles from which to draw from for their meta-analysis, a 16-fold increase. There were no journals specifically dedicated to HCS in 1999; now there are two dedicated HCS journals, and it is not uncommon to see major articles on simulation in such flagship publications as the *Journal of the American Medical Association* and the *New England Journal of Medicine*.

The advantages of this rapid growth were increased interest and support for research and implementation from healthcare institutions and foundations, federal funding sources, and industry. The disadvantages were that in the rush to implement, core concepts and guiding principles were slow to emerge, including appropriate theoretical and philosophical grounding.[1]

Of the common foundational theoretical and philosophical approaches that have emerged, the most frequently cited are experiential learning theories as described by David Kolb, Jean Lave, L. S. Vygotsky, and others. Also frequently discussed, directly or indirectly, is the educational and pragmatist philosophy of John Dewey, especially with regard to practice and incorporation of guided self-reflection (debriefing) into the simulation education experience.

In HCS education, two of the key components of the process are deliberate practice and reflection. While it is true that Dewey frequently discusses the concept of reflection—indeed in the opening paragraphs of *How We Think*, he claims that reflective thought is "the principal subject of this volume."—it is not clear that the HCS experiential learning theorists have the same thing in mind. For example, while Kolb's focus is on reflection in the moment, Dewey cites a process, occurring over time, of "active, persistent and careful consideration of any belief or supposed form of knowledge in the light of the grounds that support it, and the further conclusions to which it tends" (Dewey 1910, p. 6). I will contend here that Dewey's conception of the self-reflection and idea generation processes are misunderstood and not always congruent with those in the current HCS literature (Dewey 1910; Vygosky 1978; Schön 1983; Chickering and Gamson 1987; Lave and Wenger 1991; Kneebone 2005). The potential ramifications are that this misinterpretation clouds our understanding of HCS processes and keeps us from—advantageously—leveraging self-reflection and idea generation in ways that Dewey intended.

HEALTHCARE SIMULATION METHODOLOGY: LEVERAGING OF REFLECTION AND DELIBERATE PRACTICE

Discussions of the current conception of how learning takes place in the setting of HCS education draw heavily on Chickering and Gamson, Kolb, Lave, Schön, and to a lesser extent, Vygotsky. As mentioned above, Dewey is often cited as well. Chickering and Gamson (1987) outlined seven principles for good practice in undergraduate education: active learning, prompt feedback, student/faculty interaction, collaborative learning, high expectations, allowing for diverse learning styles, and time on task. Kolb's learning styles model (1984) drew on his own earlier work in the development of the Learning Style Inventory (LSI) (Kolb 1976, 1984). He posits that there are four types of learning styles that fall on differing parts of intersecting learning continuums (concrete to abstract and active to reflective). Lave and Wenger described their concept of situated learning as a process of legitimate peripheral participation. The emphasis is on the learner's need to "participate in communities of practitioners" and on the fact that knowledge attainment requires "full participation in the sociocultural practices of a community" (Lave and

Wenger 1991, p. 29). Donald Schön reported on his detailed observations of professional practitioners; he discovered that they all had "a capacity for reflection on their intuitive knowledge in the midst of action," what he termed reflection-in-action (1983, p. vii). Lev Vygotsky described what he believed to be a "Zone of Proximal Development" (ZPD), where learners, with expert assistance, can develop skills in problem solving (Vygotskiĭ and Kozulin 1986, Vygotsky 1978). In varying ways, all of these theorists place high value on experiential learning where both deliberate practice and reflection play prominent roles. Multiple publications in the realm of HCS have cited the work of these theorists as foundations for best practices to be included in simulation education. Two components, deliberate practice and reflection (or debriefing) in some form, are widely considered as essential to all simulation exercises, from simple to complex (Isssenberg and McGaghie 2005; Jeffries 2005).

Reflection in HCS has its roots in the recognition of its role in the learning process. Both "reflection" and "debriefing" are terms commonly found in the HCS literature. Reflection is mostly used in the broad general sense of a trainee thinking about the simulation experience, whereas debriefing typically refers to a structured reflection process guided by a facilitator in the period of time shortly after a simulation training exercise has concluded. Much of the early work in healthcare drew on the simulation training experiences of the aviation industry, where debriefing was a critical component in attempting to understand decision making in error genesis (either in simulation or in the "real world") (Waag 1988; Howard et al. 1992). Greenblat (1971, 1977) recognized the importance of debriefing as early as the 1960s in the context of computer gaming in nursing education. Ulione (1983) later advocated both for debriefing support of learners who may have experienced untoward reactions to a simulation event and in evaluating the effectiveness of the simulation itself. Robbins (1999) extended the role of debriefing to the actual clinical environment given the psychological impact of nurses working in the setting of a major disaster relief effort. Kneebone and others cite Vygotsky and his description of the ZPD as foundational underpinnings of the concept of expert-guided reflection (Kneebone 2005, p. 550; Parker and Myrick 2012). Kneebone also cites Tharp and Gallimore (1991) and Bruner's (1960) concept of "scaffolding" support by an expert facilitator. The learner first receives external help and then works toward a process of internalization. Ultimately support is withdrawn when it is no longer needed, the learner having reached some predetermined threshold of competency. The simulation literature of the last decade has also been demonstrating an emerging understanding that reflection plays an important role in facilitating retention (Brackenreg 2004; Bond et al. 2006; Dieckmann, Gaba, and Rall 2007; Fanning and Gaba 2007). Reflection in the form of a facilitator-guided debriefing is universally seen as a critical educational component, but exactly how it should be conducted is very much contested. Immediate face-to-face structured debriefing protocols appear to be superior to other methods, but even written feedback in conjunction with a screen-based trainer has been demonstrated to improve performance on follow-up standardized evaluative simulation scenarios when compared with no debriefing (Schwid et al. 2001). Internal, unguided reflection certainly occurs with the individual learner, but most of the literature has focused on the structured, expert-guided type of reflection in the form a period of debriefing post-simulation. This is now considered essential to help learners make the most of the simulation experience (Fox-Robichaud and Nimmo 2007; Kuiper et al. 2008; Owen and Sprick 2008). This author has recently proposed a new model of HCS that recognizes three distinct phases of self-reflection each capable of impacting the internalization of concepts learned in the simulation continuum. These phases are either structured or unstructured. The unstructured reflections are the least described phases of the continuum but hold much promise for target description and possible intervention.

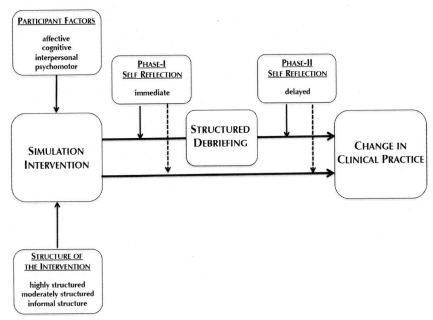

Figure 14.1   Phase I Self-Reflection (P1-SR) and Phase II Self-Reflection (P2-SR) are considered to be any reflective process done by the participant post-simulation intervention. These can be defined temporally if a structured debriefing is utilized, with Phase I occurring in the period between the simulation intervention and the structured debriefing and Phase II occurring at any time after the structured debriefing. *Source*: (Goode 2015).

Changing circumstances of day-to-day practice in healthcare have created difficulties in balancing the need for healthcare trainees to learn essential skills against the drive for patient safety. The old apprentice model of learning procedures on live patients after minimal prior preparation—the half-joking mantra was "see one, do one, teach one"—clearly doesn't align with a desire to reduce iatrogenic injuries and medical error. Donald Schön described this as being "bound to an epistemology of practice which leaves us at a loss to explain, or even to describe, the competences to which we now give overriding importance" (1983). Deliberate practice has been touted as a key to reducing errors in healthcare practice. This was highlighted in the IOM report, which specifically pointed to simulation as a means to reduce the incidence of adverse patient outcomes:

> Another example of ways to prevent and to mitigate harm is simulation training. Simulation is a training and feedback method in which learners practice tasks and processes in lifelike circumstances using models or virtual reality, with feedback from observers, other team members, and video cameras to assist improvement of skills. Simulation for modeling crisis management (e.g., when a patient goes into anaphylactic shock or a piece of equipment fails) is sometimes called "crew resource management," an analogy with airline cockpit crew simulation. (Kohn, Corrigan, and Donaldson 1999, pp. 176–177)

The promise of simulation in healthcare has been that trainees could practice in a safe environment, developing critical knowledge and psychomotor skills prior to ever touching a patient. Additionally, this training, because of the nature of the medium—a replaceable, reusable stand-in for the real thing—could be done again and again until a benchmark proficiency level was achieved. One emerging approach in HCS that leverages these advantages is mastery learning. The concept isn't new, going back at least to the work of Block and Bloom in the 1970s, but has had a rebirth with the advancement of simulation technology that theoretically allows for an infinite number of repetitions of a task or set of tasks. Barsuk, McGaghie, and Wayne describe mastery learning as a form of competency-based education that assumes uniform educational outcomes (Barsuk et al. 2010; McGaghie et al. 2012). Another way of explaining this is the substitution of a J-curve for the standard normal distribution curve for learner outcomes. The J-curve has been used to describe phenomena in areas such as economics, but it appears in the social psychology literature in the early twentieth century. Floyd Henry Allport described the J-curve as a "field of conforming behavior" with "a scale whose steps are variations of the behavior, ranging from the prescribed or 'proper' act, which most completely fulfills the behavior's purpose (on the left) to that which gives it the least recognizable amount of fulfillment (upon the right)"— a J-shaped curve of positive acceleration (Allport 1934).

Simulation-based mastery learning (SBML) proponents believe that properly designed interventions have the ability to move or accelerate learners along this curve. The variant component between learners becomes the time for each to achieve the predefined threshold performance level, but it is expected that everyone will reach that threshold. Deliberate practice, typically in a 1:1 or 1:2 instructor-to-trainee ratio, is essential for success of the method (Ericsson 2004). Using this approach, Barsuk et al. have been able to demonstrate significant retention of internal jugular central venous catheter (CVC) insertion skills at 6 and 12 months post training (decline of immediate postintervention scores of 100% to 82.4% at 6 months and 87.1% at 12 months) (Barsuk et al. 2010). Similarly, Cohen et al. were able to demonstrate that use of a simulation-based mastery learning CVC training protocol resulted in reduced catheter-related bloodstream infections and an associated decrease in hospital costs for treating such complications (Cohen et al. 2010). Because demonstrating measureable impacts of simulation training on real-world outcomes such as decreased patient morbidity and cost savings for the healthcare system has been difficult to achieve, these studies are significant. They provide strong examples of the ability of well-designed simulation educational interventions to leverage both deliberate, repetitive practice and a real-time approach to delivering a focused, structured debriefing. Current HCS educational theorists claim, by way of Kolb, Schön, and others, a link to the pragmatist philosophy of John Dewey for foundational underpinnings of reflection and deliberate practice approaches, but it may be that the linkages are not exactly as described in the HCS literature. In order to understand where these linkages fall apart, we need to introduce Dewey's philosophy of reflective thought and how it impacts education.

DEWEY ON REFLECTIVE THINKING AND THE RATIONAL ELABORATION OF AN IDEA (IDEA GENERATION)

In Dewey's classic publication *The Reflex Arc Concept in Psychology* (1896), he gives a description of the act of a child reaching for a candle, positing a continuum that formally begins in "the act of seeing; it is looking, and not a sensation of light" (Dewey 1896, pp. 358–359). The rest of the continuum follows from this as "the seeing, stimulates another act, the reaching .... More specifically, the ability of the hand to do its work will depend, either directly or indirectly, upon its control, as well as its stimulation, by the act of vision" (p. 359). For Dewey, this is an expanded sense of how an act of coordinated movement occurs; it is "an enlarged and transformed coordination; the act is seeing no less than before, *but it is now seeing-for-reaching purposes. There is still a sensori-motor circuit, one with more content or value, not a substitution of a motor response for a sensory stimulus*" (p. 359, emphasis added).

One description of reflective thinking that Dewey gives us is "that operation in which present facts suggest other facts (or truths) in such a way as to induce belief in the latter upon the ground or warrant of the former" (Dewey 1910, pp. 8–9). There is a clear parallel between his description of the sensori-motor circuit and his later descriptions of the reflective thinking process as a connected continuum, moving from the identification of a difficulty to the suggestion of an idea or explanation and ultimately the substantiated belief of a rational elaboration of an idea (for convenience I will use idea generation interchangeably with rational elaboration of an idea) (see Dewey 1910, pp. 72–78). Discrete division points in the process aren't obvious, and there are perhaps no real terminal achievements, as a frequent common "end point" for Dewey is reflection that circles back on itself in a reexamination of the initial question. This conceptualization is described in detail in Dewey's *How We Think* (1910) and again in *The Need for a Recovery of Philosophy* (1917).

Reflective thinking and idea generation are linked, but the connection is one with "more content or value," derived from the incorporation of reflection or past experience to complete the "circuit." Thought also has other important qualities such as signifying (the suggestion of something which is not observed) and belief—or disbelief—that are grounded in some form of empirical evidence, either in the present or drawn from past experience (Dewey 1910, p. 7). In *How We Think* Dewey describes a process of encountering a new situation or problem that suggests a solution. If the suggested solution is not examined, then proper idea generation hasn't occurred. "Thought is cut short uncritically; dogmatic belief, with all its attending risks, takes place. But if the meaning suggested is held *in suspense*, pending examination and inquiry, there is true judgment. We stop and think, we *de-fer* conclusion in order to *in-fer* more thoroughly. *That is to say, an idea is a meaning that is tentatively entertained, formed, and used with reference to its fitness to decide a perplexing situation, a meaning used as a tool of judgment*" (Dewey 1910, pp. 107–108). Reflective thinking that results in idea generation then is an active process based on observational and reflective activities, in much the same way that sensation is not divorced from motor response, and the analogy is used several times in *The Need for a Recovery of Philosophy* (1917):

In the orthodox position *a posteriori* and *a priori* were affairs of knowledge. But it soon becomes obvious that while there is assuredly something *a priori*—that is to say, native, unlearned, original—in human experience, that something is *not* knowledge, *but is activities made possible by means of established connections of neurons*. This empirical fact does not solve the orthodox problem; it dissolves it. (pp. 116–117, emphasis added)

Dewey would also dissolve the argument for the need to develop a theory of reality claiming that the "chief characteristic trait of the pragmatic notion of reality is precisely that no theory of Reality in general, *überhaupt*, is possible or needed. It occupies the position of *emancipated empiricism*" (emphasis added). His pragmatism is "content to take its stand with science," and ultimately it "takes its stand with daily life, which finds that such things really have to be reckoned with as they occur interwoven in the texture of events" (Dewey 1917, p. 133).

Reflective thinking plays a key role in this reckoning of events as they occur, this rational elaboration of an idea, giving us the potential to draw upon past experience, to run the data if you will, and apply the results to current situations. For Dewey, our past isn't frozen, but instead experiences are "plastic" and can be reformed to answer new and unexpected problems. And this extends to education, the role of which is to enhance what is learned experientially. "Education takes the individual while he is relatively plastic, before he has become so indurated by isolated experiences as to be rendered hopelessly empirical in his habit of mind" (Dewey 1910, p. 156). Generally, in this reflective thought-to-idea process, reflection is called upon only when conditions interrupt or seem to refute our present belief structure. That is, the process of goal attainment is blocked by an unforeseen circumstance (Dewey 1910, pp. 8–9).

DEWEY ON DELIBERATE PRACTICE

In relation to "practice," Dewey is clearly wary of modes of practice that focus strictly on the acquisition of skill at the expense of intelligent consideration through the reflective thinking process (Dewey 1910, pp. 51–52). Again, for Dewey, reflective thought isn't a discrete event; it is imbedded in the active process of how we think and there is a structure to it, "a *consequence*—a consecutive ordering in such a way that each determines the next as its proper outcome," of which some, as was pointed out earlier, reflect backward. "Each phase is a step from something to something—technically speaking, it is a term of thought. Each term leaves a deposit which is utilized in the next term" (pp. 2–3). Both an initial inductive reflection and then a deductive reflection processes are used in the reflective thinking construct as a whole. In this way, we use our memory, our specific store of empirical data, to create more general rules and constructs—idea generation. For Dewey, "A complete act of thought involves both—it involves, that is, a fruitful interaction of observed (or recollected) particular considerations and of inclusive and far-reaching (general) meanings" (pp. 79–80).

And so, even when learning is focused on skill acquisition, "there is danger of the isolation of intellectual activity from the ordinary affairs of life." He believed that this was true of "logical studies" as well as of training in skill acquisition where expediency in the teaching-learning environment often "makes the subjects *mechanical* and thus restrictive of intellectual power" (pp. 50–51). A significant question for our modern HCS educational structure is whether we have fallen into this epistemological trap with regard to reflection and practice.

CONGRUENCE AND DISAGREEMENT BETWEEN DEWEY'S PRAGMATIST PHILOSOPHY AND CURRENT HEALTHCARE SIMULATION PRACTICES

The early phase of the current HCS movement was very much focused on psychomotor tasks. There was a strong appeal to grounding HCS in educational theories that emphasized the importance of doing things, either as a primary learning mode or as supplemental and supportive of classroom learning. The Kolb learning styles model, with its intersecting learning continuums of active to reflective and concrete to abstract, seemed ideally suited to describe the role of simulation activities (Sandmire, Vroman, and Saunders 2000; Hauer, Straub, and Wolf 2005).[2] Likewise for situated learning theory with its emphasis on knowledge accumulation through the performance of a variety of activities that are embedded in the culture of a practice community, in this case the community of healthcare practitioners (Clancey 1995; Kneebone 2005). Action and doing were and continue to be listed as key strengths of HCS education, and it is easy to find supporting evidence in John Dewey's version of pragmatism:

> The entire scientific history of humanity demonstrates that the conditions for complete mental activity will not be obtained till adequate provision is made for the carrying on of activities that actually modify physical conditions, and that books, pictures, and even objects that are passively observed but not manipulated do not furnish the provision required. (Dewey 1910, p. 100)

Here is a definite call for experiential learning, and the introduction of affordable high-fidelity mannequins and part-task training devices seemed to be tools well suited to answering it by removing passivity and engaging the learner in an active, lifelike setting. We should, however, be wary of superficial similarities when looking for theoretical and philosophical grounding. And à la Dewey we need to guard against conditioning influences that might impact the method of instruction. He lists these as "(1) the mental attitudes and habits with whom the child is in contact; (2) the subjects studied; (3) current educational aims and ideals" (Dewey 1910, pp. 46–47). We will address them where applicable below. Earlier, we explored Dewey's concepts of reflective thought and practice; these are good places to search for congruence with current thought and practice in HCS education.

Practice

Dewey's views on education are often mentioned as the theoretical and philosophical underpinnings for newer theories developed to describe the role of both practice and reflection. The work of David Kolb and Donald Schön are good (but not the only) examples. Frequently, it appears that the referencing is out of context and Dewey's more nuanced modeling of thought generation is condensed down to the simple mantra of "learning by doing." For example, in a discussion of Tolstoy's interest in education, Schön states, "His own school anticipated John Dewey's later approach to learning by doing" (1983, pp. 65–67). There are some similarities as Tolstoy, like Dewey, believed that instruction should be individualized to the student, but this statement is an oversimplification at best. These types of distillations of Dewey's thinking have infused the HCS educational literature, at times resulting in a large emphasis on psychomotor practice divorced from the relevant didactic and clinical practice grounding. In *How We Think*, Dewy expresses serious concerns that, for the sake of educational expediency and efficiency, "the pupil is enjoined to do this and that specific thing, with no knowledge of any reason … his mistakes are pointed out and corrected for him; he is kept at pure repetition for certain acts till they become automatic" (Dewey 1910, p. 51). In the immediate period after the 1999 IOM report, efforts to leverage simulation seemed to be following just this pattern. It was an obvious approach to use given that one of the main advantages of simulation lies in the ability for the trainee to repeat the same process with the same level of fidelity for as many times as seems necessary to attain the desired learner outcome. Linkages to curricular content or contextually grounding references to the clinical setting were often lacking. Dewey derisively equates this repetitive drilling approach to one "conceived after the mechanical analogy of driving, by unremitting blows, a foreign substance into a resistant material" (1910, p. 63). Pure repetitive drilling runs the risk of falling under the second adverse conditioning influence, misclassifying the mode of teaching for the subject studied. As we discussed in the previous section, task repetition divorced from its contextual application can be considered as detrimental to the development of reflective thought. In one sense, to be truly effective, practice must be like the reflex arc example described earlier. The psychomotor must be contextualized and guided by the sensory input, and this in turn is tempered by past experience. Earlier we mentioned SBML as a promising approach in HCS education, and it seems to meet these requirements.

On the surface, SBML seems to be drilling by purposeful design as often it is applied to a psychomotor task, but here the task practice is highly grounded in clinical context and there is usually a requirement to achieve a baseline knowledge threshold prior to entering the simulation setting. The assumption of a J-curve for all learner outcomes as opposed to the usually assumed normal or Gaussian distribution is critical; that is, it is assumed that all learners can master the expected performance goals (McGaghie et al. 2009). Another key difference is that no set number of repetitions required is assumed; the instructor approach is to do "whatever it takes to achieve the desired threshold competency." The work cited earlier by Barsuk, McGaghie, and Wayne demonstrates knowledge and skill gains that have translated to clinical practice with reductions in measureable markers such as central line–related infection rates and healthcare costs (Barsuck, Cohen et al. 2009; Barsuk, McGaghie, Cohen et al. 2009; Cohen et al. 2010). So here, the approach appears to have good overlap with Dewey's conceptualization. The instructor needs to assure that the ability of the learner to "fasten upon and single out (abstract, analyze) those features of one experience which are logically best" isn't "hindered by premature insistence upon their explicit formulation. It is repeated use that gives a method definiteness; and given this definiteness, precipitation into formulated statement should follow naturally" (Dewey 1910, p. 113).

Recent literature around mastery learning in HCS has also revealed how the discipline must learn to address Dewey's first potential adverse conditioning influence: "A teacher's best conscious efforts may be more than counteracted by the influence of personal traits which he is unaware of or regards as unimportant" (1910, p. 47). The evidence is strong that simulation educational interventions are equal or superior to all other traditional forms of instruction; Barsuk et al. have demonstrated this for nurses and intensive care unit residents, an outcome that was not unexpected based on earlier results from the HCS literature (Barsuk, McGaghie, Cohen, Balachandran, et al. 2009; Barsuk, McGaghie, Cohen, O'Leary, et al. 2009; Barusk et al. 2014; Barsuk et al. 2015). What was unexpected were the findings regarding those who were training the trainees. These studies and others reveal that, based on referencing to standardized and validated training protocols, the instructors also did not meet threshold benchmarks at baseline (Clark et al. 2014; McQuillan et al. 2015). One assumption is that trained, practicing professionals might deviate from evidenced-based practice guidelines based on their own judgment and experience. Whether these variant individual choices in practice patterns are better or worse remains an open question.

The end goal of mastery learning—and, of course, of HCS in general—is to provide a base of experience, gained away from live patients, which will allow for safe practice in the clinical setting. It is also hoped that these simulated experiences will allow the learner to generate new solutions within the bounds of learned best practices when faced with unfamiliar patient presentations in the real-world setting. Not all of this can be achieved through the practice piece alone, whether simulated or real; this is where the power of reflection in idea generation is intended to be realized.

Reflection

As we have seen earlier, "reflection" is embedded in the continuum of Dewey's conception of what thinking is, not a disconnected activity, which is nicely demonstrated in a passages from *How We Think*:

> Given a difficulty, the next step is suggestion of some way out-the formation of some tentative plan or project .... The data at hand cannot supply the solution; they can only suggest it. What, then, are the sources of the suggestion? Clearly past experiences and prior knowledge. (Dewey 1910, p. 12)

It is now widely accepted that reflection in the form of a structured debriefing should be a component of all simulations, simple or complex (Issenberg et al. 2005; Jeffries 2005; McGaghie et al. 2010). Vygotsky's Zone of Proximal Development is considered by some in the simulation community to be foundational to the concept of expert-guided reflection (Kneebone 2005; Parker and Myrick 2012). Recall that Kneebone (himself referencing Tharp) describes a process of the learner first being dependent on external help (the instructor) to grasp the significance of the simulation experience. In current HCS educational approaches, this takes the form of the structured debriefing. In the context of the simulation educational event, this role for reflection fits Dewey's model. Reflection is part of the educational experience and hopefully part of the learner's continuum of idea generation. One area of concern is that after the simulation experience, the trainee continues to reflect, this time without the aid of an expert facilitator. Earlier a newly proposed model of HCS was introduced (Goode 2015). This model identifies three distinct phases of reflection, one immediately after the simulation event, but prior to the structured debriefing (Phase I Self-Reflection, or PI-SR), the debriefing and then a longer phase of self-reflection post-simulation (Phase II Self-Reflection, or PII-SR). We know almost nothing about these unguided self-reflection periods and how trainees then internalize their simulation experience into their mental model of the task or process that the simulation was directed at. It may be, as Dewey quips about a child experiencing a dog, that the learner "begins with whatever significance he has got out of the one dog he has seen, heard, and handled" (1910, p. 128). In many educational settings, one dog may be all the trainee gets!

It is possible, though, that with thoughtful curricular design, learners may be able to take away with them enough information to sustain reflective thought and rational elaboration of new ideas in the face of similar, but slightly variant, circumstances to that which they were exposed in simulation. This type of model for the formation of new ideas and solutions to process problems from prior experience is supported by the work of Chi et al. (1989), who examined the self-generated explanations of students studying presolved mechanics problems. A principle finding was "that 'good' students learn with understanding: they generate many explanations" that "result in example-independent knowledge." All of this sounds remarkably like Dewey's contention that "conceptions are general because of their general use and application, not because of their ingredients. The view of the origin of conception in an impossible sort of analysis has as its counterpart the idea that the conception is made up out of all the like elements that remain after dissection of a number of individuals. Not so; the moment a meaning is gained, it is a working tool of further apprehensions, an instrument of understanding other things. Thereby the meaning is *extended* to cover them .... Synthesis is not a matter of mechanical addition, but of application of something discovered in one case to bring other cases into line" (1910, p. 129).

This conception can inform our thinking about the larger goal of simulation educational efforts: transference (or translation) of simulation learning to the clinical setting, thereby improving patient outcomes. While growing, the literature is relatively sparse for demonstration of transference (Cook et al. 2011). However, figure 14.1 allows us to see that transference is not a singular, indefinable event that happens at a discrete point post-simulation. Transference represents a process or series of processes that are posited to result in a new conceptualization of the targeted task. This new conceptualization subsequently allows the learner to change clinical practice patterns in a positive way, in the manner of Dewey's reflective thinking continuum. A theoretical model that lays out the pieces of this continuum in HCS education and places them in a temporal context can provide understanding of where gaps in our current understanding exist.

The endpoint of independent novel idea generation, based on prior experience, is very much in line with the goals in healthcare education of developing "critical thinking," which is believed to occur as the trainee moves along a continuum from "novice-to-expert." The novice-to-expert process is described by Patricia Benner's model of the same name, where proficiency develops across a spectrum of five levels of practice proficiency. Benner never addressed concrete steps for the trainee to achieve expert-level practice skills, nor has she ever addressed the potential role for HCS in this process, but her model heavily implies gaining and drawing on past experiences in a way that sounds quite like Dewey's conception of attaining expertise; partially it is derived from innate qualities of each individual, "but it also represents the funded outcome of long familiarity with like operations in the past. Possession of this ability to seize what is evidential or significant and to let the rest go is the mark of the expert, the connoisseur, the judge, in any matter" (1910, p. 104).

While it is not explicitly mentioned, Benner of course implies the need for critical thinking skills to be gained in the journey across the novice-to-expert spectrum, a journey that almost certainly involves self-reflection on experiences and input from mentors. Kolb and Schön would contend that experts reflect in the moment. Although definitive evidence has yet to emerge as to its efficacy in this regard, HCS has increasingly been mentioned as a means of helping trainees to develop critical thinking skills. A significant part of the power of simulation to help foster the development of critical thinking *may be the encouragement of reflection on the simulation experience*, and this most likely occurs in both a structured and an unstructured manner.

As alluded to earlier, much is drawn from recent theories of adult education to provide theoretical support for simulation. David Kolb's theory of experiential learning is one of the most frequently cited. While experiential learning no doubt plays a significant role in the simulation educational experience, we should be cautious in drawing parallels between these theories and Dewey's (or others) pragmatist philosophy. Kolb, for example, claims that his experiential learning cycle draws significantly from Dewey's philosophy and he gives a summary of Dewey in his text (1984, pp. 22–23). Reijo Miettien has criticized Kolb's interpretation as being both incorrect and misleading. He concludes that "Kolb does not give an adequate interpretation of Dewey's concept of experience and reflective thought. Kolb speaks about experiential learning. Dewey speaks about experimental thought and activity. These terms are phonetically close. However, they are theoretically and epistemologically quite far apart" (Miettenen 2000, p. 70). For example, while Kolb's focus is on reflection in the moment, Dewey cites a process, occurring over time, that will "rest upon the sound assumption that observation is an active process. Observation is exploration, inquiry for the sake of discovering something previously hidden and unknown, this something being needed in order to reach some end, practical or theoretical … observation proper is searching and deliberate" (1910, p. 193).

SUMMARY

In summary, while much of the HCS literature claims to draw on John Dewey's conceptualizations of reflective thought and educational practices, there are often disconnects between the source material and its translation to the current simulation practice. Misinterpretation of grounding principles and philosophy can result in falling into the traps of the conditioning influences that Dewey identifies in *How We Think*—influence of the habits of others, the nature of the subject being studied, and the influence of current societal aims and ideals. If we consider it of value, and we should, there are good examples in the current HCS literature of approaches remaining true to Dewey's version of the pragmatist philosophy. His pragmatism "takes its stand with daily life, which finds that such things really have to be reckoned with as they occur interwoven in the texture of events" (Dewey 1917, pp. 116–117). Martin Schiavenato (2009) and others have pointed out that many of those using HCS have failed to ask the basic question of what defines something as simulation, why are we using it, with whom, how are we deploying it and when. Given that this emerging science has yet to clearly define its own theoretical and philosophical underpinnings, working toward an understanding of the continuum of HCS grounded by a solid understanding of Dewey's conceptualizations of the processes of thought and idea generation is a fine place to start.

NOTES

1. As of this writing, only one widely accepted theoretical model of HCS processes has been proposed. I will discuss here ongoing work for my own proposed theoretical model.
2. These continuums are also sometimes referred to as processing information (watching to doing) and perception (feeling to thinking).

REFERENCES

Allport, Floyd Henry. 1934. "The J-Curve Hypothesis of Conforming Behavior." *Journal of Social Psychology* 5:141–83.

Barsuk, Jeffrey H., Elaine R. Cohen, Joe Feinglass, William C. McGaghie, and Diane B. Wayne. 2009. "Use of Simulation-Based Education to Reduce Catheter-Related Bloodstream Infections." *Archives of Internal Medicine* 169 (15):1420–3.

Barsuk, Jeffrey H., Elaine R. Cohen, William C. McGaghie, and Diane B. Wayne. 2010. "Long-term Retention of Central Venous Catheter Insertion Skills after Simulation-based Mastery Learning." *Academic Medicine* 85 (10 Suppl):S9–12.

Barsuk, Jeffrey H., Elaine R. Cohen, Anessa Mikolajczak, Stephanie Seburn, Maureen Slade, and Diane B. Wayne. 2015. "Simulation-Based Mastery Learning Improves Central Line Maintenance Skills of ICU Nurses." *Journal of Nursing Administration* 45 (10):511–7.

Barsuk, Jeffrey H., William C. McGaghie, Elaine R. Cohen, Jayshankar S. Balachandran, and Diane B. Wayne. 2009. "Use of Simulation-Based Mastery Learning to Improve the Quality of Central Venous Catheter Placement in a Medical Intensive Care Unit." *Journal of Hospital Medicine* 4 (7):397–403.

Barsuk, Jeffrey H., William C. McGaghie, Elaine R. Cohen, Kevin J. O'Leary, and D. B. Wayne. 2009. "Simulation-Based Mastery Learning Reduces Complications

During Central Venous Catheter Insertion in a Medical Intensive Care Unit." *Critical Care Medicine* 37 (10):2697–701.

Barsuk, Jeffrey H, Elaine R. Cohen, Steven Potts, Hany Demo, Shanu Gupta, Joe Feinglass, William C. McGaghie, and Diane B Wayne. 2014. "Dissemination of a Simulation-Based Mastery Learning Intervention Reduces Central Line-Associated Bloodstream Infections." *BMJ Quality & Safety* 23 (9):749–56.

Bond, William F., Lynn M. Deitrick, Mary Eberhardt, Gavin C. Barr, Bryan G. Kane, Charles C. Worrilow, Darryl C. Arnold, and Pat Croskerry. 2006. "Cognitive Versus Technical Debriefing after Simulation Training." *Academic Emergency Medicine* 13 (3):276–83.

Brackenreg, Jenni. 2004. "Issues in Reflection and Debriefing: How Nurse Educators Structure Experiential Activities." *Nurse Education in Practice* 4 (4):264–70.

Bruner, Jerome S. 1960. *The process of education.* Cambridge,: Harvard University Press.

Chi, Michelen T. H., Miriam Bassok, Matthew W. Lewis, Peter Reimann, and Robert Glaser. 1989. "Self -Explanations: How Students Study and Use Examples in Learning to Solve Problems." *Cognitive Science* 13:145–82.

Chickering, Arthur W., and Zelda F. Gamson. 1987. "Seven Principles for Good Practice in Undergraduate Education." *The Wingspread Journal* 9 (2).

Clancey, William J. 1995. "A Tutorial On Situated Learning." *International Conference on Computers and Education*, Taiwan.

Clark, Edward G., James J. Paparello, Diane B. Wayne, Cedric Edwards, Stephanie Hoar, Rory McQuillan, Michael E. Schachter, and Jeffrey H. Barsuk. 2014. "Use of a National Continuing Medical Education Meeting to Provide Simulation-Based Training in Temporary Hemodialysis Catheter Insertion Skills: A Pre-test Post-test Study." *Canadian Journal of Kidney Health and Disease* 1:25.

Cohen, Elaine R., Joe Feinglass, Jeffrey H. Barsuk, Cynthia Barnard, Anne O'Donnell, William C. McGaghie, and Diane B. Wayne. 2010. "Cost Savings from Reduced Catheter-Related Bloodstream Infection after Simulation-Based Education for Residents in a Medical Intensive Care Unit." *Simulation in Healthcare* 5 (2):98–102.

Cook, David A., Rose Hatala, Ryan Brydges, Benjamin Zendejas, Jason H. Szostek, Amy T. Wang, Patricia J. Erwin, and Stanley J. Hamstra. 2011. "Technology-Enhanced Simulation for Health Professions Education: A Systematic Review and Meta-Analysis." *JAMA* 306 (9):978–88.

Dewey, John. 1896. "The Reflex Arc concept in Psychology." *Psychological Review* 3 (4):357–70.

Dewey, John. 1910. *How We Think.* Boston: DC Heath & Co.

Dewey, John. 1917. "The Need for a Recovery of Philosophy." In *The Pragmatism Reader: From Peirce through the Present*, edited by Robert B. Talisse and Scott F. Aikin, 109–140. Princeton, NJ: Princeton University Press.

Dieckmann, Peter, David Gaba, and Marcus Rall. 2007. "Deepening the Theoretical Foundations of Patient Simulation as Social Practice. [see comment]." *Simulation in Healthcare: The Journal of the Society for Medical Simulation* 2 (3):183–93.

Ericsson, K. Aanders 2004. "Deliberate Practice and the Acquisition and Maintenance of Expert Performance in Medicine and Related Domains." *Academic Medicine* 79 (10 Suppl):S70–81.

Fanning, Ruth M., and David M. Gaba. 2007. "The Role of Debriefing in Simulation-Based Learning." *Simulation in Healthcare: The Journal of the Society for Medical Simulation* 2 (2):115–25.

Fox-Robichaud, Alison E., and Graham R. Nimmo. 2007. "Education and Simulation Techniques for Improving Reliability of Care." *Current Opinion in Critical Care* 13 (6):737–41.

Goode, Joseph S. 2015. *Finding Our Way to a New Model of Healthcare Simulation* paper presented at the 15th Annual International Meeting on Simulation in Healthcare. New Orleans, LA.

Greenblat, Cathy S. 1977. "Gaming-Simulation and Health Education an Overview." *Health Education Monographs* 5 (suppl 1):5–10.

Greenblat, Cathy S. 1971. "Simulation Games and the Sociologist." *American Sociologist* 6 (2):161–64.

Hauer, Patrick, Christina Straub, and Steven Wolf. 2005. "Learning Styles of Allied Health Students Using Kolb's LSI-IIa." *Journal of Allied Health* 34 (3):177–82.

Howard, S. K., D. M. Gaba, K. J. Fish, G. Yang, and F. H. Sarnquist. 1992. "Anesthesia Crisis Resource Management Training: Teaching Anesthesiologists to Handle Critical Incidents." *Aviation Space & Environmental Medicine* 63 (9):763–70.

Issenberg, S. Barry, and William C. McGaghie. 2005. "Features and Uses of High-Fidelity Medical Simulations that can Lead to Effective Learning: A BEME Systematic Review." *Medical Teacher* 27 (1):1–36.

Issenberg, S. Barry, William C. McGaghie, Emil R. Petrusa, David L. Gordon, and Ross J. Scalese. 2005. "Features and Uses of High-Fidelity Medical Simulations that Lead to Effective Learning: A BEME Systematic Review." *Medical Teacher* 27 (1):10–28.

Jeffries, Pamela R. 2005. "A Framework for Designing, Implementing, and Evaluating Simulations used as Teaching Strategies in Nursing." *Nursing Education Perspectives* 26 (2):96–103.

Kneebone, Roger 2005. "Evaluating Clinical Simulations for Learning Procedural Skills: A Theory-Based Approach." *Academic Medicine* 80 (6):549–53.

Kohn, Linda T., Janet M. Corrigan, and Molla S. Donaldson. 1999. *To Err is Human: Building a Safer Health System*. Washington, DC: National Academy Press.

Kolb, David A. 1984. *Experiential Learning: Experience as the Source of Learning and Development*. first ed. Englewood Cliffs, NJ: Prentice Hall. Original edition, 1984.

Kolb, David A. 1976. *Learning Style Inventory: Self-scoring Test and Interpretation Booklet*. Boston: McBer.

Kuiper, RuthAnne, Carol Heinrich, April Matthias, Meki J. Graham, and Lorna Bell-Kotwall. 2008. "Debriefing with the OPT Model of Clinical Reasoning During High Fidelity Patient Simulation." *International Journal of Nursing Education Scholarship* 5 (1):1–13.

Joseph S. Goode

Lave, Jean, and Étienne. Wenger. 1991. *Situated Learning: Legitimate Peripheral Participation.* Cambridge University Press.

McGaghie, William C., S. Barry. Issenberg, Emil R. Petrusa, and Ross J. Scalese. 2010. "A Critical Review of Simulation-Based Medical Education Research: 2003–2009." *Medical Education* 44 (1):50–63.

McGaghie, William C., S. Barry Issenberg, Elaine R. Cohen, Jeffrey H. Barsuk, and Diane B. Wayne. 2012. "Translational Educational Research: A Necessity for Effective Health-Care Improvement." *Chest* 142 (5):1097–103.

McGaghie, William C., Viva J. Siddall, Paul E. Mazmanian, Janet Myers, Health American College of Chest Physicians, and Committee Science Policy. 2009. "Lessons for Continuing Medical Education from Simulation Research in Undergraduate and Graduate Medical Education: Effectiveness of Continuing Medical Education: American College of Chest Physicians Evidence-Based Educational Guidelines." *Chest* 135 (3 Suppl):62S–68S.

McQuillan, Rory F., Edward Clark, Alireza Zahirieh, Elaine R. Cohen, James J. Paparello, Diane B. Wayne, and Jeffrey H. Barsuk. 2015. "Performance of Temporary Hemodialysis Catheter Insertion by Nephrology Fellows and Attending Nephrologists." *Clinical Journal of the American Society of Nephrology.* 10 (10):1767–772.

Miettinen, Reijo. 2000. "The Concept of Experiential Learning and John Dewey's Theory of Reflective Thought and Action." *International Journal of Lifelong Education* 19 (1):54–72.

Owen, Harry, and Cyle D. Sprick. 2008. "Simulation Debriefing and Quantitative Analysis using Video Analysis Software." *Studies in Health Technology & Informatics* 132:345–7.

Parker, Brian C., and Florence Myrick. 2012. "The Pedagogical Ebb and Flow of Human Patient Simulation: Empowering Through a Process of Fading Support." *Journal of Nursing Education* 51 (7):365–72.

Robbins, Ian. 1999. "The Psychological Impact of Working in Emergencies and the Role of Debriefing." *Journal of Clinical Nursing* (3):263–8.

Sandmire, David A., Kerryellen G. Vroman, and Ronda Sanders. 2000. "The Influence of Learning Styles on Collaborative Performances of Allied Health Students in a Clinical Exercise." *Journal of Allied Health* 29 (3):143–9.

Schiavenato, Martin. 2009. "Reevaluating Simulation in Nursing Education: Beyond the Human Patient Aimulator." *Journal of Nursing Education* 48 (7):388–94.

Schön, Donald A. 1983. *The Reflective Practitioner: How Professionals Think In Action.* New York: Basic Books.

Schwid, Howard A., Geroge A. Rooke, Piotr Michalowski, and Brian K. Ross. 2001. "Screen-Based Anesthesia Simulation with Sebriefing Improves Performance in a Mannequin-Based Anesthesia Simulator." *Teaching & Learning in Medicine* 13 (2):92–6.

Thorp, Roland and Ronald Gallimore. 1991. "A Theory of Teaching as Assisted Performance." In *Child Development in Social Context 2: Learning to Think*, edited by Paul Light, Sue Sheldon and Martin Woodhead. London and New York: Routledge.

Ulione, Margaret S. 1983. "Simulation Gaming in Nursing Education." *Journal of Nursing Education* 22 (8):349–51.

Vygotskiĭ, Lev S., and Alex Kozulin. 1986. *Thought and Language*. Translation newly rev. and edited / ed. Cambridge, MA: MIT Press.

Vygotsky, Lev S. 1978. *Mind in Society*. Boston: Harvard University Press.

Waag, W. L. 1988. "Programs and Prospects in Aircrew Performance Measurement." *Aviat Space Environ Med* 59 (11 Pt 2):A46–51.

Index

About the Editor and Contributors

EDITOR

Robyn Bluhm is an Associate Professor in the Department of Philosophy and Lyman Briggs College at Michigan State University, USA.

CONTRIBUTORS

Rachel A. Ankeny is a Professor in the School of Humanities at the University of Adelaide, Australia.

Sophie van Baalen is a PhD Candidate at the University of Twente, The Netherlands.

Suze Berkhout is a Resident in the Department of Psychiatry at the University of Toronto, Canada.

Marianne Boenink is an Assistant Professor in the Department of Philosophy at the University of Twente, The Netherlands.

Mieke Boon is a full Professor in Philosophy of Science in Practice at the University of Twente, The Netherlands.

Anthony Vincent Fernandez is a Killam Postdoctoral Fellow in the Philosophy Department at Dalhousie University, Canada.

Tania Gergel is a Visiting Research Fellow in Philosophy and Psychiatry at King's College London, UK.

Maya J. Goldenberg is an Associate Professor in the Department of Philosophy and the Bachelor of Arts and Science (BAS) program at the University of Guelph, Canada.

Joseph S. Goode is a Clinical Instructor in the School of Nursing at the University of Pittsburgh, USA.

Susan C. C. Hawthorne is an Associate Professor in the Department of Philosophy at St. Catherine's University, St. Paul MN, USA.

Thomas Kabir leads the Public Involvement in Research (PIiR) program at The McPin Foundation in the UK.

James Krueger is an Associate Professor in the Department of Philosophy at the University of Redlands, USA.

Keekok Lee is an Honorary Research Professor/Fellow in the Faculty of Humanities at the University of Manchester, UK.

Christopher McCron is a student at the University of Guelph, Canada.

Delphine Olivier is a PhD student at the Université Paris 1 Panthéon-Sorbonne/IHPST, France.

Abraham P. Schwab is an Associate Professor in the Department of Philosophy, Indiana University—Purdue University Fort Wayne, USA.

Mark R. Tonelli is a Professor in the Division of Pulmonary and Critical Care Medicine at the University of Washington, USA.